THE SOLAR ELECTRIC HOME

a photovoltaics how-to handbook

Joel Davidson
Richard Komp

aaTec

Third Printing 1984

Copyright © 1983 by **aatec publications**
PO Box 7119, Ann Arbor, Michigan 48107

Library of Congress Catalog Card Number 83-70647
ISBN 0-937948-04-7

All Rights Reserved
Manufactured in the United States of America
Printed and Bound by McNaughton & Gunn, Inc., Saline, Michigan

Distributed to the trade by
Kampmann & Company, Inc., New York

Illustrations by Lawrence Komp
Cover Design by Carl Benkert

PREFACE

This book was written to fulfill a need expressed by many—
"Where can I find a good book that will tell me step-by-step how
to use solar electric power?" *The Solar Electric Home* gives you
the information to determine your power requirements, size your
PV power system, make intelligent decisions about energy production and consumption, and install your own system. It also
directs you to other information sources and suppliers. It is hoped
that this book, and its companion *Practical Photovoltaics*, will
serve to advance our common goal of energy self-reliance.

Dedicated to the memory of

DAN TRIVICH

a pioneer in the solar cell field

—RJK

ABOUT THE AUTHORS

Joel Davidson has been using solar electricity since 1978. He has lectured extensively about solar energy and has taught carpentry and appropriate technology. He founded the PV Network and *The PV Network News,* helping hundreds of people get started in photovoltaics. As a special consultant, he brings a comprehensive approach to energy problem-solving.

Richard J. Komp, PhD, physicist/chemist, designed and built Skyheat, a nonprofit solar research complex in southern Indiana, where he conducts solar cell research and holds hands-on PV workshops. In 1982, Dr. Komp co-founded SunWatt Corporation which developed and now manufactures hybrid PV systems. He is the author of *Practical Photovoltaics: Electricity from Solar Cells.*

ACKNOWLEDGMENTS

There are many people who deserve thanks: Val Bertoia, who makes beautiful windchargers; Steve Willey, who has been an inspiration to many solar/wind users; Paul Wilkins and Fran Foster of Solar Works! who were always there when I needed them—all deserve my deepest gratitude. I also want to thank the people of the PV Network who have shared their experiences. Very special thanks go to Bill and Marge Lamb and the William Lamb Company.

—JD

I wish to thank Steve Cook for teaching me to combine solar cells and wind generators; Carol Harlow, Jim Welch, and the many others who helped organize PV workshops; and Susan Phillips, Sis Kaiser, and Shiela Seewald who assisted in the development of the first hybrid and other photovoltaic modules.

—RJK

CONTENTS

CHAPTER 1
INTRODUCTION TO PHOTOVOLTAICS

To put the magnitude of the sun's power into perspective, at noon the solar energy striking an area 70 miles long by 70 miles wide, if converted into photovoltaic electricity, would equal the peak capacity of all the Earth's existing power plants. A solar cell power plant covering only 1% of the Sahara Desert would produce all the electricity consumed on this planet.

But to put this global data into down-home language, the surface of the average roof, if covered with solar cells, could provide for the needs of an all-electric home with every possible modern device, including heating and air-conditioning. What makes this even more amazing is that solar electricity for home use is practical right now for the energy-conservative do-it-yourselfer.

We now know that solar heating is less costly than conventional heating in most instances. Many heat their homes and work places, cook, distill water and dry crops with the sun. The sun captured in greenhouses extends the normal growing season, increasing food production. In industry, solar furnaces are now used for process heat applications. Solar energy has proved itself in these applications. Now, sunlight can be converted directly to electricity by using solar cells.

1

ELECTRICAL POWER SOURCES—
THE OPTIONS

When we think about the lights, motors, pumps and other electrical devices we use every day, it is easy to see how electricity has become one of the major building blocks for modern society. Unfortunately, electricity requires special means of production, and in general that means the consumption of nonrenewable resources like coal, oil and gas. These raw materials are being depleted at a rapid rate and within the next few generations our reserves may be consumed.

Nuclear energy is being used for the production of electricity, but recent controversy has clouded its practicality and safety. It is generally believed that nuclear power plants are dangerous. Indeed, the disposal of spent radioactive materials alone poses such a serious threat to life that it makes other forms of waste and pollution seem relatively insignificant. Research into fusion energy (thermonuclear power) has not yet produced a safe, practical alternative to the "dirty" technology of nuclear power production.

Wind, geothermal and water power are safe and sane solutions to the nuclear power problem. Unfortunately, these natural resources are not everywhere and at all times available. It has been estimated that hydroelectric technology could produce 25% of the U.S. electrical needs, but other nations are not sufficiently endowed with major mountain ranges to provide adequate sites for hydroelectric plants. Some natural geothermal sites are being tapped for electricity production, but current locations and methods are too limited to compete with fossil fuels as an energy source. Ocean or tidal power generation still remains on the drawing board. Wind power, once a major contributor to the energy picture before rural electrification, will become more widely used but is limited to sites where the installation of a windcharger will produce a net energy gain.

The subject of energy production and consumption is a complex issue with many political, economic and social implications for both present and future generations. As fossil fuels are depleted, the scramble for the few remaining pockets of conventional fuels will continue to destabilize the world political and economic situation, leading to ever greater conflict and confrontation. A virtually limitless and globally available source of energy like solar electricity offers a solution.

THE POLITICAL SITUATION

There has been a dramatic increase in government and private research reports that solar electric power for residential energy consumption is just a few years away. Rather than encourage the actual use of residential photovoltaics, these announcements inhibit consumer acceptance. Recent news stories are making everyone familiar with the potential for solar electricity. Unfortunately, the emphasis on potential rather than present-day usage has done more to slow the widespread use of solar cells than their present high cost. Again and again, we hear that new developments in solar cell technology will make it possible for us to use solar cells in the future.

Because of the projected 1986 government goal of low-cost photovoltaics (70 cents per watt), many prospective users have decided to wait before converting to solar electric power. Reports of existing expensive and complicated residential systems scare away those who do not want to wait. If a transitional approach toward photovoltaic power for homes were encouraged by government and industry, more people could participate in current technology, thereby providing the needed capital and consumer acceptance to the fledgling photovoltaics industry.

The perception that residential photovoltaics is not practical *now* is only part of the problem. The practitioners of residential photovoltaics along with much small-scale or appropriate technology are viewed at best skeptically by the mainstream scientific and business communities and at worst as modern-day Luddites who threaten existing systems and institutions. Until the people who are working with residential photovoltaics and other small-scale technologies are recognized and until they have credibility with the mainstream of the business and technological community, residential photovoltaics will affect only the fringes of our society.

The negative reaction to this small-scale individual approach to providing for our electrical needs is, in part, a reaction against technology that is consciously or subconsciously viewed as a threat to the status quo. Extensive use of photovoltaics would decentralize power generation and generally lessen individual dependence on centralized manufacturing and distribution systems.

The basic assumption that solar cells are too expensive is misleading. Actually, solar cells are a practical way of producing electricity for many uses. But present attempts to power all-electric homes with photovoltaics are defeatist due to the very high electrical demands of some appliances and equipment and the current high cost of residential solar cell modules. However, low-wattage, solid-state equipment, energy-efficient appliances, and the use of 12-volt DC recreational vehicle equipment together with an assortment of other energy inputs attuned to local availability can bring down electrical demands. Battery storage, inverters and synchronous inverters make autonomous and/or grid-connect systems practical.

There are several reasons for what seems to be footdragging on the part of industry, the government, utility companies and the public. Recently, with the threat that Japan may take over the photovoltaics market, American manufacturers have begun to wake up. There are numerous examples of either mismanagement or direct attempts to hold back public use of solar electricity, but the monopolistic nature of the solar cell industry itself is also to blame. Vast amounts of capital are currently required to go into cell production, and only large companies like oil corporations have that kind of money. Obviously, oil companies are not going to give whole-hearted support to a product that competes with oil profits.

Of course, industry has accused the government of failing to help solar cell development. During the Carter administration this trend was changed, but the Reagan administration has turned away from government-funded solar research. The government-supported research that has survived has placed more emphasis on space applications than on domestic or terrestrial use of solar cells.

Another monopoly has a vested interest in solar cells. The utilities and coal, oil and gas companies, although they have been working with solar cells, are worried about how solar cells will be used. The power and profits of these companies come from their control of centralized energy production plants. Uniquely, solar cells installed at the point of consumption could replace the large factories, plants and energy production units that now provide our electricity. It is easy to see that the possibility of eliminating the need for power plants and their associated power lines, equipment and personnel poses a real threat to those in control of energy production.

But "the big boys" are not entirely to blame. Americans' love for gadgets and luxury coupled with little knowledge about energy are among the main reasons why solar cells are not in common use. Even with the recent attention given to energy, most Americans do not know enough about energy and electricity to make intelligent decisions about its production and use. Rather than opening up the vast store of knowledge available, the government, industry, utility companies, the media and the average citizen have created more myth and mystery.

When asked, the average American perceives only two things about electricity—power companies produce it and it costs too much. Most Americans never think about the electricity they use. If they did, they would begin to see that the equipment and appliances they use are energy-inefficient and wasteful. Not only are the electrical devices we use wasteful, but our patterns of use are wasteful. Almost everyone every day leaves lights and equipment on and operating when not needed. This behavior and our "bargain basement" attitude about buying energy-conserving equipment are prime reasons why solar cells are not in widespread use today. As long as we fail to assume the responsibility of learning the basics of power usage and production, we must accept high utility bills.

On a positive note, the rate of growth in the power consumed in this country is leveling off—we are using less electricity than predicted a few years ago. The power companies are having trouble justifying the completion of new power plants planned and started five years ago when the utility industry expected a rosy future of ever-increasing per capita electricity consumption. As the prices of photovoltaic systems drop and as more people realize that they can lead comfortable lives without being tied to the utility grid, more and more roofs will display the solar arrays that tell the passerby, "Here lives someone who has taken a step toward energy independence."

ENERGY CREDITS AND TAX REFUNDS

As of this writing, there has been talk around Washington to repeal the federal energy credits. Unless that happens, we have a good law on the books. For years taxpayers have been subsidizing nonrenewable energy businesses. Worse yet, we are forced to pay millions of dollars each year to prop up a failing nuclear power industry.

It is important for all of us to know about *and use* the federal energy tax credits. These savings alone can cut 40% from the initial cost of a PV home power system. In addition, almost all states have some form of energy credit, refund or rebate program to further reduce initial costs. These credits can mean savings as high as 80%.

One reason given by the Administration for eliminating the energy credits is that very few people are using them. In 1979 nearly 5 million Americans spent over $3 billion on energy and conservation and, using the tax credits allowable under the law, each saved an average of $100—a total of almost $500,000,000 stayed in the pockets of these energy-conscious people.

Rather than list the various state rebates, refunds and credits, we strongly encourage you to find out more about them. Write to your senators and congresspeople and let them know that the tax credits for renewable energy systems are important to you. Let your opinions be known.

So far we have been able to keep the federal information hot-line alive with our support. For information about all aspects of solar and conservation and whom to contact in your state, call the Conservation and Renewable Energy Inquiry and Referral Service 1-800-523-2929 (Puerto Rico and the Virgin Islands 1-800-462-4983; Alaska and Hawaii 1-800-523-4700). Or write Renewable Energy Information, P.O. Box 8900, Silver Spring, MD 20907.

GRID-CONNECT VERSUS THE AUTONOMOUS HOME

Federal law now requires utility companies to buy power from small producers. Many people think that they can set up a wind or solar electric system and sell excess power to the utility company and use the grid instead of batteries to keep system costs down. If it were that simple, it would be great. However, upon examination, it is not as good as it seems. Grid-connect PV homes and a buy/sell relationship with the utility company may not be the best way to go. Here are some factors to consider before making the grid-connect commitment.

Sure, the utility company will buy your excess power. But, remember, they sell to you at retail and buy back at wholesale. The idea of running your meter backwards sounds good, but the power companies will not permit it. They will install two meters— one for you and one for them. In addition, the synchronous inverter or power conditioning equipment required to interface with the utility company is very expensive. The special inverter requires utility company approval and often costs over $3000. A utility engineer or your engineer, if approved by them, must be on hand during installation of the interface equipment, increasing costs even more.

But let's say you have the money and the desire to install a large PV array and grid-connect system. Along comes a utility company power failure. You think that your own system will operate because the sun is shining—but no. The synchronous inverter must get power from the grid to operate. That means you are out of power even if the wind is turning your charger and the sun is shining. Thus, you still need a battery bank and possibly another inverter for grid power outages.

We recommend that people who are considering a grid-connect system gradually move in that direction. First, install a stand-alone or autonomous PV system with batteries and then convert some of your equipment to low-power DC or use an inverter. After you have had a taste of this independence, then decide if you still want to cogenerate with the grid. If you do, you can always use the initially installed equipment in the event of a power company blackout or brownout.

In the future, large-scale use of photovoltaic electricity in this country will undoubtedly be accomplished using grid-connect systems. The economy-of-scale in storage systems and the maintenance involved in battery storage systems will encourage the majority of users to remain attached to the grid after their solar cell roof shingles are installed.

PHOTOVOLTAICS—A TECHNICAL OVERVIEW

Ever since 1839 when the French physicist Edmond Becquerel discovered that copper oxide electrodes in a liquid could produce an electric current when struck by light, the direct conversion of sunlight into electricity has been an exciting goal. Charles Fritts, who made the first selenium photovoltaic cells in the 1880s, predicted that buildings with roofs covered with solar cell arrays would generate their own electricity. The intense interest in photovoltaics waned, however, when it became impossible to produce amorphous selenium solar cells with even 1% efficiency. Except for a small amount of work on cuprous oxide cells in the

1930s, this interest was revived only when solid-state researchers at Bell Labs developed the single-crystal silicon cell in the 1950s. At 15% efficiency, this cell turned out to be the perfect electric power generator for the fledgling space satellites. Simple to use and extremely dependable, the single-crystal silicon cell could operate indefinitely on the continuously available sunlight in space.

At one time, solar cells that did not meet the strict requirements of the space program were the only ones available for terrestrial use. Now manufacture has expanded to meet some of our daily energy demands. These almost-magic electric power sources have again captured the imagination of those who seek energy independence.

WHAT SOLAR CELLS ARE AND HOW THEY WORK

Photovoltaic theory states that radiated light energy either direct from the sun or diffused through the atmosphere or from an artificial source such as a light bulb consists of a stream of energy units (photons). These photons strike the solar cell and create an electron flow (electrical energy) in the cell. This can be understood if you think of the solar cell as containing an internal barrier, where photons give up their energy to an electron that jumps the barrier.

Most commercial solar cells are made of crystalline silicon, one of the most abundant elements on the planet. It is a constituent of sand, but the sand used as the starting material for silicon solar cells is very different from the sand we see at the beach. The silicon used for solar cells is very pure and does not conduct electricity. It is called a semiconductor. Pure silicon has few free electrons and is a good insulator, whereas copper with a whole spectrum of free electrons is a good conductor. By doping the silicon crystals with elements that have different numbers of electrons (such as boron or phosphorus), the material is made more conductive to electricity. When phosphorus is added, each phosphorus atom contributes an additional free electron, making what is called an n-type semiconductor. If boron atoms are added to the crystal, each one has one less electron than do the silicon atoms, and each can create a "hole" (a place where an electron should be, but isn't) making a p-type semiconductor.

To better understand how solar cells work, it is helpful to know how they are made. A large crystal of pure silicon containing a tiny bit of boron is "grown" under exacting laboratory conditions of heat and vacuum. The resultant crystal is sliced into extremely thin wafers, which are then treated in a diffusion furnace so that phosphorus impurity atoms are added in a thin layer at the top of each wafer. The zone between the p-type material in the bulk of the wafer and the n-type surface layer creates the barrier, called a p–n junction.

When light energy—photons—is absorbed in the doped wafer, negative and positive charges are created in the solar cell. Since the doping has created dissimilar properties in the wafer, this charge can flow as an electrical current in only one direction. The output voltage produced across the cell under full-sun conditions is determined by the height of the internal barrier. By connecting wires to each side of the wafer, the current produced can be used in an electrical current. For those of you who desire more information, *Practical Photovoltaics* contains a detailed explanation of how solar cells are made and how they work.

NEW PRODUCTION TECHNIQUES

The production of solar cells is complex and requires much skilled labor, and the sophisticated equipment used requires high capital investments. The need for hand labor, special equipment and conditions, the waste of the wafer cutting process, and the growing shortage of high-grade silicon all add to the cost of solar cells. The exacting conditions used throughout the manufacturing process also escalate costs.

However, mass production of solar cells by automated equipment puts them within the budget of many people. Most manufacturers are geared to the exacting standards of the space industry which requires cells of a reliability and dependability not needed in terrestrial cells. Hopefully, more manufacturers will respond to the growing market for residential solar equipment and produce a dependable, low-cost product.

We are often asked, "Shouldn't I wait until solar cells are cheaper? I've just read about this major breakthrough. . . ." Solar cells are expensive today and it would be disappointing to buy a high-priced system only to find that you could have purchased the equivalent amount of power for one-fifth the price just by waiting a couple of months. But that's not going to happen, and here are some of the reasons.

Figure 1.1 shows the U.S. Department of Energy projections for solar cell prices (in dollars per watt) for the next few years. We have added the best PV module prices now available to an individual, based on bulk purchase. The actual price of solar cells fell annually until 1980 when inflation outdistanced technical development. The price picture currently is artificial and complicated by the involvement of oil companies but, at least for the present, the price seems to have stabilized at around $8 to $10 per watt. Any reductions in price for crystalline solar cell modules will be minor and will occur slowly over the next few years.

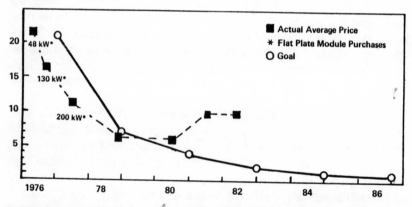

FIGURE 1.1—Cost of photovoltaics: history and goals. For several years, solar cell module prices were actually lower than DOE goals, but recently prices have gone up. DOE no longer sets price goals.   [Courtesy: DOE]

Real price breakthroughs will require radical departures in the production process of silicon solar cells. Researchers are working on new ways to produce silicon wafers from cubes of lesser-grade materials and on ribbon growth methods to produce rectangular solar cells with few processing steps. Commercial modules produced via these two avenues are already available, but ironically they are slightly more expensive than those made by more conventional techniques. Breakthroughs have occurred, but much remains to be done before they will be reflected in silicon cell price reductions.

Amorphous silicon cells, under development for more than a decade, promise low-price PV systems, but their record so far has been one of high costs and questionable long-term stability. The Japanese are now producing millions of amorphous silicon cells for pocket calculators and digital watches. These cells degrade quickly when exposed to full sunlight, but intensive research is underway to remedy that defect. It is impossible to predict when reliable low-cost amorphous power modules will be commercially available.

Currently available cadmium sulfide/cuprous sulfide cells similarly degrade, but from exposure to moisture in the air. SES (a Shell subsidiary) is selling cadmium sulfide/cuprous sulfide modules hermetically sealed in glass cases. These extremely reliable devices are as expensive as silicon cells, but recently Photon Power Corporation (jointly owned by a French oil company and Libbey-Owens-Ford) announced for the third time the impending availability of CdS/Cu_2S modules at a price of $5 per watt. By the time these devices are in the hands of consumers—with aluminum cases replacing the wooden ones—the price in units of one will probably be only slightly less than present silicon solar cell modules.

Prices for solar cells will come down, but it's going to take a long time. Photovoltaic systems are practical, dependable and affordable now—why wait?

A practical application . . . .

PV system on a farm in Carter County, Missouri. The system consists of used Spectrolab panels generating 312 peak watts, Gould deep-cycle batteries with 1100 ampere-hour storage, and a Silicon Sensors, Inc., 50-amp voltage regulator. This system was designed and built by Environmentally Appropriate Technology, Inc., Fremont, Missouri.
[Photo: D. P. Grimmer]

CHAPTER 2
LIVING WITH PV

ENERGY CONSERVATION

It cannot be stressed too often—PV systems work well *if you are energy-conscious.* You must be willing to keep an open mind and find creative ways to use and *not* use electricity.

We are just beginning to pay the real cost for energy. (Here we are talking present and projected monetary costs, but we should also keep in mind that the real cost of electricity includes the environmental and societal impacts that will be borne by future generations.) Electric rates are rising so rapidly that utility companies have submitted multiple rate hikes for approval. It is realistic to assume that these increases are just the beginning of substantial electricity price increases. If your $47 per month electric bill is subject to a modest 12% yearly rate increase, in 20 years your monthly bill will be over $350! When you figure what electricity has cost you, is costing you, and will cost you, PV power system costs are not that high.

The first step a person or household must take when deciding whether or not to set up a PV power system is to stop using luxury "toys" like electric can openers and blow driers. For some people these devices are important, and they must pay for using them. But, for most of us, our lives would be pleasantly simplified if we went back to basics and let our fingers do the working. No one likes to be told what can or can't be done in one's own home, but as brownouts, blackouts and rationing increase, the time will come when the choice will not be ours.

Right now, we still can make some choices. A number of new appliances are designed to be energy-efficient. Some, including air conditioners and refrigerators, bear government-mandated efficiency tags so you can intelligently compare different brands. It will almost always pay in the long run to purchase the most energy-efficient brand available, even if the alternative is less expensive. Some appliances made in the 1950s are actually more energy-efficient and reliable than those manufactured today. If you are looking for a used refrigerator, the 20-year-old manual defrost type may give you better service than a 5-year-old fancy chrome and avocado side-by-side model. Of course, you will have to lug that messy tray of water and look at rounded corners.

On the other hand, many electrical devices in common household use were designed for the days of subsidized "cheap" energy. Take a good look at the electrical equipment in your home. Are you still using an old TV or stereo? Check the identification plate (usually on the back) and see if the device uses more than 100 watts. If so, it is probably a relic from the days of low-cost grid power. Check all your devices and appliances. You'll be surprised to find that it is those cookers and heaters that really eat up the watts. The domestic electric water heater is one of the most wasteful appliances, unless it has a timer to shut it off when not in use. Air conditioning (which is really air refrigeration) and cooking with electricity (except brief periods of microwave use) may one day be outlawed unless the user either produces his/her own power or is willing to pay a penalty charge for high consumption. Home heating with electricity is so expensive that it is second only to air conditioning in power consumption in the temperate zones.

Incandescent lights (the ones with filaments that heat) are big wasters of electricity. A good way to check if a light is efficient is to touch it when it is on. Those old bulbs in your home are better suited for heating than for lighting. Fluorescent lamps are more energy-efficient, and DC fluorescents do not have the noticeable 60-cycle flicker that AC lamps have. A 20-watt fluorescent bulb will produce the same illumination as an 80-watt incandescent lamp—and that's real energy savings.

Just as important as converting to more efficient devices and appliances is the need to change our energy consumption patterns. If you are the kind of person who leaves lights on in unoccupied rooms or uses more lights than necessary, you will have to develop different energy habits before you can change over to low-cost home-grown electricity. Even if you do not produce your own energy, it is wise to become aware of your energy usage and conserve. Most homeowners could cut their electrical consumption in half without suffering. The children of the future will appreciate your conservation efforts.

In Europe, where high utility bills have been a way of life for decades, smaller, efficient appliances are standard. Europeans are becoming almost as enamored with gadgets as we are, but their whole-hearted acceptance is tempered by a desire to remain independent of their "labor-saving" devices. Well-designed, unobstrusive blenders, hair driers and dish washers seem appropriate in the smaller apartments that are the norm and usually consume less than half the wattage of their American counterparts. Scandinavians, Germans and the Swiss enjoy a higher standard of living than we do while using half the energy per capita.

Some European appliances are beginning to show up here, notably the demand hot water heaters. These make excellent point-of-use backup systems for solar water heater installations. One problem with European appliances is that they are generally designed to be used on the 230-volt/50-Hz current that is the utility standard there. Many, however, are convertible with the simple flick of a built-in switch. European incandescent 230-volt light bulbs have finer filaments and are not as efficient as our 120-volt bulbs. The lower the design voltage of an incandescent bulb, the more efficient it will be.

* * *

This is a good time to put in a plug for one of the most valuable books we have read since installing our own PV power systems. Michael Hackleman, a long-time user of wind and solar power and a tinkerer who knows how important conservation is to the future, has written a great book called *Better Use of . . . Your Electric Lights, Home Appliances, Shop Tools—Everything That Uses Electricity* (1981, available from Earthmind, 4844 Hirsch Road, Mariposa, CA 95338). For years we have been telling people to stop buying books, use that money for hardware and get out there and "do it." But Michael's book has changed that viewpoint. Now we say, "Read Michael's book first." It will tell you how to use the electrical equipment you want to use properly with your home power system: It will answer many questions about modification of existing equipment for more efficiency. It might frustrate the unhandy among us, but it will make it possible for those who are all thumbs to be able to explain what they want done. The only thing wrong with this book is that it was not available when we were first setting up home energy systems.

TWO PV SYSTEMS

PV systems are as varied as the people who install and use them. Each of us has different needs, different amounts of cash to work with, and different ways of expressing or satisfying our needs. A plumber might install a solar water heater while a carpenter would build an attached solar greenhouse. A mechanic might build a windcharger while someone with less mechanical experience would be inclined to put in a PV system. While one person might tinker with his/her home power system, another person would want it installed and done with.

PV power systems require some do-it-yourself work even if you hire someone to do the initial installation. Each installation has its story. The following are descriptions of our PV systems and their evolution. Hopefully, yours will be more straightforward as you learn from our experiences.

Joel Davidson's Arkansas Homestead

In 1972 my wife, daughter and I left the city to live in the Ozarks in northwest Arkansas. After renting and leasing for a few years, we moved atop a 2200-ft-high mountain in spring 1976. The nearest power line was one mile away. We had the general idea of producing our own electricity as part of our desire to eliminate monthly bills paid to insensitive monopolies and to gain greater control over our lives. Now we were faced with the reality of the project. We talked and visited with others and concluded that we lived at a good wind site and eventually would install a windcharger as money became available. Things were tight and other

FIGURE 2.1–Joel Davidson's Pettigrew, Arkansas, residence. This small PV system uses only four 35-watt ARCO Solar modules and a 420 ampere-hour battery bank in the cabin crawlspace to power lights, TV, stereo, radio, fans, computer, appliances, tools and more.   [Photo: Joel Davidson]

needs were more pressing, like home and barn building and just day-to-day living.

For shelter we quickly erected a pole-frame cabin with south-facing windows. The sun provided 50% of our daytime winter heat, but wood from our land was the main heat source. We installed our propane stove and refrigerator and used kerosene lamps for lighting. After drilling a dry hole instead of a well, we were forced to haul water. Finally, we installed a 500-gallon cistern and a hand pump. The tank, too small even for our modest needs, convinced us that a sure cistern is better than a chancy well-drilling experience. Even if we had hit water while drilling, we would have had the problem of getting the water out of the well.

We built and tested several low-cost (under $50) solar water heaters, but none could withstand the quick freezes typical of this region, so we opted for a used 30-gallon propane water heater for only $17. We installed the water heater keeping the pilot light off when not in use. Using a spark igniter to light the water heater and cook stove was inconvenient but really saved on fuel. In our first year we used less than 130 gallons of propane to cook, heat water and refrigerate.

We began our electrical system with a couple of fluorescent lamps (16 watts) purchased from J. C. Whitney. Power came from a truck battery which was charged by daily trips to and from work. We bought a 12-inch black and white TV set at that time. Battery power worked well except over long weekends in winter. Power consumption would deep-discharge the battery and the truck wouldn't start Monday mornings.

A friend of ours had a windcharger he wasn't using, so we borrowed it. Before we got everything hooked up, the windcharger's blades, in an attempt at self-destruction, gyroscoped into the tower in gusting winds. That mechanical failure and our studies about solar cells told us to spend our savings for our first PV power set-up. So by Christmas 1978, as a result of the combination of windcharger failure, a good deal from Free Energy Systems (FES) and our desire to be PV pioneers, we were watching news reports about snowstorm power outages on our solar-powered TV.

Wiring in our cabin was simple, consisting of secondhand outlets and scavenged wire and plugs. We had no regulator and used a standard automobile battery. It was pretty primitive—but it worked. The wire was No. 12/2 Romex run the shortest distance possible to keep resistance losses to a minimum. The outlets and plugs were marked with red paint to indicate polarity.

In the beginning we didn't have much sun power. The three 10-watt modules from FES which made up our array were mounted on a piece of plywood which could be tilted up or down to adjust to the seasonal angle of the sun. We operated our 16-watt fluorescent lamps 2 to 4 hours per day and the TV about the same amount of time. We bought a 12-volt DC water pump from Airborne Sales and installed it beneath the cabin, tapping a pipe into the cistern. That gave us running water inside the cabin, drawing only 4.6 amps when operating. For music we used a Radio Shack am/fm cassette stereo with a home-made pre-amp (made from an IC chip), and an old turntable retrofitted with an external 12-volt DC motor and belt drive.

Production from the PV array in June 1979 matched our consumption—a very modest 3 kilowatt-hours per month. In winter we used more lighting because of shorter days, more cloud cover, and more time spent indoors. Winter PV production was only one-fourth of consumption, but that was no problem because we had upgraded our battery storage. We bought some used deep-cycle batteries at salvage prices that were in like-new condition. Six heavy 2-volt batteries were hooked up in series to give us 12 volts and over 400 ampere-hours of storage. The batteries are rated for an 8-hour discharge at 52.5 amps and are sold by C&D Batteries. We got ours from our windcharger friend, who got them from the telephone company.

Because our consumption was greater than our production at times, and as it is a good idea to keep batteries "topped off," fully charged to ensure long life, we assembled a standby gasoline generator. We put together a 3.5-hp Briggs and Stratton engine from an old tiller (horizontal shaft) and a used truck generator and regulator. To connect the two shafts we used a flex-coupling. The whole thing was bolted to a piece of truck frame and put in a dog house-type shelter away from the cabin. Heavy wire purchased by the pound from a salvage company ran from the backup generator to the batteries, which were located in the house. We adjusted the regulator for a 30-amp charge. In 18 months we used only 8 gallons of gasoline. The charger would run for 6 hours and for 180 ampere-hours of battery storage (6 times 30 amps equals 180 ampere-hours) on three-fourths of a gallon of gasoline.

By this time we were accustomed to our power system and things were getting more comfortable. We had a shower installed but still used an outhouse. In summer we had fans for cooling and plenty of power to run them. Also, by this time we were totally committed to PV power and were saving our money to buy more modules. By participating in a bulk purchase organized with Steve Cook at North Arkansas Community College, we were able to get a good price on four 33-watt ARCO PV modules. Thus, we became 100% solar-powered and eliminated the need for the backup generator. We also got an ARCO Village Power Panel, which is a combination regulator and junction and breaker box. It was a big step, but we took it thinking that we might buy even more modules with our 40% federal energy credit on that purchase.

During that time we were diverting money from our home building project into our power system and other needs. We decided to sell our 32 acres on the mountain and buy 5 acres from friends in a valley nearby. We built a cabin on our new land and moved. We even moved our power system. As it turned out, the move convinced us that PV power is portable. We moved everything in one trip, removing the PV array from the roof and disassembling the modules, which fit into the cab of our pick-up truck. The batteries, which weighed over 800 pounds, were tied into the back of the truck to prevent spillage. We even took out the wiring and fixtures and reused them in our new home. In a few days we had everything hooked up again. We could have done the move and reinstallation in one day.

At our new cabin we fastened the four ARCO modules to the roof in a fixed mount. The batteries were placed beneath the cabin on a pallet. Wiring came from the batteries into the cabin and then to the ARCO Village Power Panel. Lights and outlets were wired in a few hours. The FES array was set up on the ground to charge an automotive battery used as a portable 12-volt DC power source.

That portable power pack (just a battery and carrying strap) came in handy for pumping water from a nearby shallow well. We bought a 550-watt Tripp-Lite inverter from J. C. Whitney and were able to run our power drill and saber saw outside. The inverter came indoors and was wired to the battery bank to power our 120-volt AC sewing machine, vacuum cleaner, and a small massager to ease those aching muscles. Moving had really gotten to us by now and we were ready to settle down.

By summer 1981 we had organized more PV bulk purchases, helping dozens of others around the country get started with photovoltaics. We laid out our new house site and were saving money to hire a backhoe and put in the foundation. But an excellent deal came along and once again we spent most of our savings for four used ARCO PV modules that would double our production.

PV power is cheaper than utility grid power in many cases right now. *If* you live one-half mile or more from power lines, and *if* you can do-it-yourself, and *if* you are very energy conservative, and *if* you can master the simple skills needed to wire DC circuits, then the initial cost of solar electricity is less than having utility power brought in.

Richard Komp's Skyheat Complex

My experience with solar cells is, in many ways, the exact opposite of Joel's. I started by doing basic research on new, inexpensive types of photovoltaic cells, but it was 15 years before I had the opportunity to use them in any practical way.

I grew up in Park Forest, Illinois, a community noted for its early interest in solar homes. I picked Wayne State University to do graduate work because a professor there, Dan Trivich, was developing new ways to convert sunlight directly into electricity. The idea of using the sun for power was very intriguing and the chance to work in photovoltaics, a field just then receiving some recognition, was not to be passed up.

During the course of four years investigating the "Photovoltaic Emission from Single-Crystal Cuprous Oxide," I managed to make a number of solar cells and study their properties. The idea, still a valid one, was to make solar cells in a simple manner from very inexpensive starting materials. We envisioned photovoltaic "shingles" that would cost little more than ordinary roofing materials but would furnish all the power needed by the building they covered.

After earning my PhD and finding there were no jobs related to solar, I took a position with Xerox's research labs working on new copying systems. They weren't at all interested in solar energy, but at least I got to work on the interaction of light with solid surfaces. During the course of this research, I discovered an organic semiconductor dye system that would produce a photovoltaic effect. Later, after becoming a physics professor at Western Kentucky University, I investigated this pigment in binder system, which had the potential of being the ultimate in low-cost solar cells; you simply painted it onto the substrate. (A solar cell in a spray can?) During this period in the early 1970s, I managed to build my first solar home (all passive) and get myself fired for (among other complaints) "teaching students to question established values." For awhile, I took a job fixing copying machines.

Finally able to purchase land and move to the country with several friends, in 1975 we started to construct the facilities in southern Indiana that were to become the Skyheat research center. The building ended up a hybrid structure utilizing both active and passive solar heating and a good deal of passive cooling (needed, but not sufficient in the hot, sticky weather that is the summer norm in the Ohio Valley).

In fall 1975, I was invited to join a photovoltaics research project back at Wayne State University. The Energy Research and Development Administration (ERDA), later to be called the Department of Energy, had been formed and the government was willing, at long last, to put money into photovoltaics research. It was good to be back in the mainstream of the work that was my greatest interest. It was now possible, with a real research budget, to undertake experiments on the cutting edge of the field and to

FIGURE 2.2–Richard Komp's Skyheat complex near English, Indiana. This structure, while still connected to the utility grid, serves as a test-bed for new photovoltaic ideas and systems.   [Skyheat photo]

meet and interact with the other researchers at universities, government labs, and the small companies that were making and marketing solar cells.

In addition to the hundreds of experimental cuprous oxide cells we made in the lab, I was able to acquire some high-quality conventional solar cells. My first practical project was a solar battery charger for "Entropy," my brother's small sailboat. This array lasted for about two years before it failed, but it proved a valuable lesson in how to connect solar cells and utilize them in a rather demanding environment.

The second practical PV project in Indiana was for a neighbor, Raymond Frady, who lives down the road from Skyheat. He was complaining about the high cost of kerosene, so for his 72nd birthday, we gave him his first electric light, powered by an experimental solar cell array. This system was the prototype of the hybrid PV hot water system I was later to develop to a commercially usable device.

Beth Hagens and Jim Laucus, two friends from Governor's State University near Chicago, heard about Raymond Frady's light and persuaded me to give a series of PV workshops, which they assisted in organizing. These were the first Skyheat workshops, a successful project that continues, in expanded form, to this day.

After a series of government-funded projects, including several trips to Europe to do research and demonstrate how to make the new, inexpensive solar cells, I decided in 1980 to leave Wayne State permanently and survive on my own at Skyheat. This was even before the Reagan administration delivered its devastating blows to alternative energy projects. Things have been somewhat hand-to-mouth, but the trickle of money from workshops and consulting has made it possible to keep up a small amount of research activity. Most has been development work on the types of solar cell modules assembled by the workshop participants. It is now possible for workshop attendees to construct solar cell arrays that are as efficient and professional looking as those purchased commercially. My first book, *Practical Photovoltaics*, started as an outgrowth of workshop notes but ended up a much more comprehensive volume.

In the fall of 1981, a colleague in Colorado, Carol S. Harlow, and I determined that the systems I had been working on were sufficiently developed, so we started a new photovoltaics company called SunWatt. We have available a set of rugged solar modules intended for marine and mobile use, and have just completed the design and final development work on the first commercially available hybrid PV/solar hot water system. The SunWatt H-150 module can furnish all the electrical and hot water needs for a remote homesite.

At the present time, I am busy converting Skyheat to a 100% photovoltaics-powered structure (using the solar cell modules that have collected during the course of the experimental work of the past few years). From the beginning, solar energy and wood have furnished all the heat the building requires. The system has been improved over the years with the addition of more passive elements and the recent installation of a wood-fired boiler that feeds hot water to the pipes buried in the concrete floors. Right now, all of Skyheat's hot water needs and about half the electricity are provided by the sun. Many of the lights, a small refrigerator, the water pumps, a few power tools (like soldering irons and an electric drill), a small computer, and even the telephone answering machine are powered by the 12-volt DC system.

Future plans include work on wind generators capable of utilizing the low wind speed in the area, and we expect to be electrically self-sufficient soon. I have already arranged with the chief engineer of the local rural electric co-op to run some cogeneration experiments, so we will continue to be connected to the grid as long as they are willing to work with us.

As resources permit, future Skyheat research will include alternative refrigeration systems, still a problem in the self-sufficient household.

CHAPTER 3
MAKING THE DECISION

If you are in the process of making the decision of whether or not to use solar cells to power your home, there are three basic considerations you are probably contemplating: 1. What are the advantages and disadvantages of solar cells? 2. How much is the system likely to cost? 3. Do I have a suitable location on or near my home to install the system? This chapter briefly treats these questions.

To help you decide whether or not photovoltaic power is suitable for you, we have compiled a list of advantages and disadvantages.

ADVANTAGES

1. There is a one-time cash outlay to purchase solar cells.
2. Federal and state energy credits reduce initial cash outlay 40% or more.
3. There is no monthly utility bill.
4. Users are not affected by electricity price increases or inflation.
5. The modules are reliable, sturdy and lightweight.

6. They can be used anywhere the sun shines.
7. There are no moving parts to wear out or break.
8. Modular system design can be augmented as money permits and needs require.
9. DC appliances are compatible with recreational vehicle equipment.
10. Modules can be used in conjunction with commercial electricity or wind or water power.
11. Components can be combined as a hybrid system with solar heat collector.
12. Users are not affected by commercial power outages.
13. PV does not pollute at point of use.

DISADVANTAGES

1. The initial purchase price for solar cells is high.
2. Electricity is not produced at night or on very cloudy days.
3. Storage batteries must be serviced.
4. An inverter must be used to power standard AC appliances.
5. The power output per dollar invested is low.
6. A backup generator or other power source may be required to keep batteries fully charged.
7. The manufacture of solar cells may produce some environmental pollution.
8. Small solar cell arrays should be adjusted seasonally to receive maximum sunlight.

HOW MUCH DOES A PV SYSTEM COST?

One of the main reasons solar cells are not in widespread use today is cost. Compared to utility company electricity, solar electricity is expensive. But, in the long run, a photovoltaic system may prove less expensive than commercial electricity. As electric rates continue to rise, the time will soon come when solar electricity will be a bargain.

Example 1

The cost of a typical PV 35-watt starter kit made of new components is listed:

1 each solar cell module (ARCO 16-2000)	$375
1 each Sears PV/marine Diehard battery	80
1 each Sears 12-volt fluorescent lamp	35
50 feet of No. 2 wire No. 10 gauge conductor	20
1 set of battery terminal connectors, fuses, plugs, etc.	10

$520
approximate 1982 prices
before tax credits

FIGURE 3.1–Photowatt starter kit: 35-watt module, deep-cycle battery, fluorescent lamp and wire.   [Photo: Photowatt, International]

As you can see, the solar cell panel is the highest priced part of the starter kit. Why? Production costs, although lower than ever before, are still high. However, it must be remembered that the initial cost of the solar panel is a one-time cost; the panel will last indefinitely. Compared to the monthly electric bill, which regularly increases, solar cells begin to look attractive. Many of us would already be energy-independent if we could have put all the money we have paid for commercial electricity over the past 20 years into buying solar cells. Like home insulation, solar cells will pay for themselves over a period of time. A major cost consideration with all alternative energy systems is the federal tax credits of 40% plus the available state rebates, credits and refunds. In most cases, this important subsidy can cut system costs in half, thus cutting pay-back time in half, too.

Example 2

A photovoltaic module can be combined with a solar water heater collector. Such a device, called a hybrid system, costs only slightly more than a simple solar cell array of equivalent electrical wattage. This single installation can save the homeowner a great deal of money compared to two separate systems. (Hybrid systems are covered in more detail in Chapter 12.) The cost of a 150-watt hybrid PV hot water system is as follows:

Hybrid PV/hot water solar collector (SunWatt H-150)	$2100
Batteries, 500 ampere-hour, 12-volt	450
Regulator control panel	144
Outlets, wire, switches, etc.	150
Built-in 12-volt, fluorescent lamps	150
12-volt DC circulator pump*	127
12-volt draindown valve*	93
Storage tank and plumbing*	200
12-volt DC pressure water pump*	435
550-watt Tripp-Lite inverter	170

$4019

*For solar hot water and plumbing systems   approximate 1982 prices before tax credits

The above prices are list prices. Using some of the sources given in Appendix G or information in *The PV Network News* can save you a great deal. If the price sounds out-of-range, consider that, with tax credits, the cost of being independent of the electric company will be only $2400 (less, if you can take advantage of state credits). This is considerably less than the cost of running a power line one-half mile or more to a remote homesite and enables a person following the "living lightly" philosophy to have a comfortable, convenient lifestyle. Remember, the system described gives you hot and cold running water as well as enough electricity to operate a wide range of appliances.

These examples are for illustration only. In the design section of the book, we give you the tools to custom-assemble your own system.

SITE SELECTION, TILT AND
TRACKING CONSIDERATIONS

Obviously, the site of your PV array must be where it can get full sunlight for the full day. If your home is in the woods surrounded by trees which provide shade (passive cooling), don't start cutting down trees. Spend some time sitting and thinking on the south side of your home. Consider the summer cooling and beauty of the trees as opposed to the harsh sun and stumps. Look at the sun path and note when shade covers your proposed location. It may be that some judicious trimming of a few branches or merely moving the array a few feet will solve the problem.

Although the winter sun is lower in the sky and casts longer shadows, deciduous trees with open branch patterns (and without their leaves) sometimes let enough sunlight through. In the South, where cooling is the greater problem, it is best to leave as much vegetation around the house as possible. In this case, you may have to locate the array away from the house—on the lawn, near the garden, or on a pole mount.

Selecting the site for your PV array can become overly complicated. Let's make it simple by first saying that your array should have full sunlight striking it between the hours of 9:00 a.m. and at least 3:00 p.m. Early morning sunlight is diffused because of atmospheric moisture or haze. For fixed mounts it is difficult to catch the late-summer sun after 4:00 p.m. without sacrificing ease in mounting.

Solar angles (altitude and azimuth) define the shade pattern or solar window. The solar window is the area of the sky through which the sun appears to travel (see Figure 3.2). This is important because solar collectors must be in the sun to work. As mentioned above, during the optimum solar radiation collection hours from 9:00 a.m. to 3:00 p.m., the solar window must be free of shading by trees and buildings. The Energy Task Force of New York suggests a way to estimate the solar window. Stand where the collector is to be mounted and face true south. Point so that your finger and your eye are horizontal. As shown in Figure 3.3, place one fist on top of the other (the exact number of times to be determined by consulting the table.) Sight over top of fist at true south and 30° east and west (with adjustments in fist height) to determine shading effects. Any objects above your fists will cast a shadow on the collector. Anything below your fists will be below the lowest path of the winter sun.

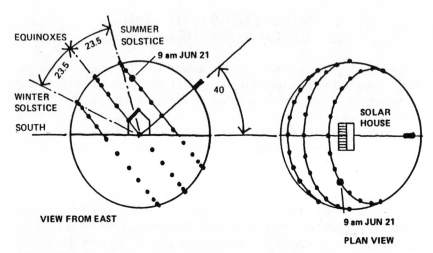

FIGURE 3.2a—If the sun's path were traced on a spherical ceiling, the view
from the east and from the top would look like this. Each day's path
describes a circle around the axis pointing toward the north, the axis here
tilted from the horizon by a latitude angle of 40°.   [Courtesy: HUD]

FIGURE 3.2b—Another way to plot the same data
is via a Mercator sky map which shows the sun's
path on a rectangular grid of azimuth and
elevation. Clear plastic guides are available from which you can make a working version of this diagram.
[Courtesy: HUD]

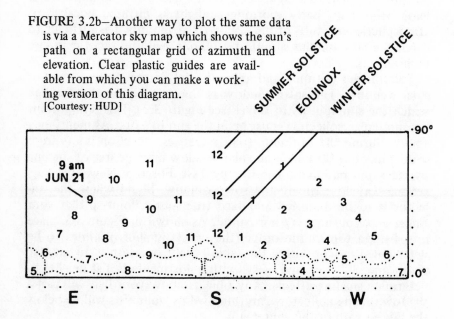

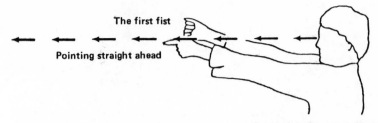

The first fist

Pointing straight ahead

Latitude	12 O'Clock Position = 0° Bearing	11 O'Clock Position (East) and 1 O'Clock Position (West) = 30° Bearing Angle
28°N	4½ fists (47° Alt.)	3 fists (30° Alt.)
32°N	3½ fists (34° Alt.)	2½ fists (26° Alt.)
36°N	3 fists (30° Alt.)	2¼ fists (23° Alt.)
40°N	2½ fists (27° Alt.)	2 fists (20° Alt.)
44°N	2¼ fists (23° Alt.)	1½ fists (17° Alt.)
48°N	2 fists (20° Alt.)	1½ fists (14° Alt.)

FIGURE 3.3—Determining the solar window or shade pattern.
[Courtesy: The Energy Task Force of New York]

It goes without saying that in the northern hemisphere your array should face south—as close to true south as possible. Magnetic south, or compass south, is not the same as true south. Check Figure 3.4 or ask a local surveyor for the magnetic declination of your area. However, don't worry too much about this as your array can deviate as much as 15° from true south and still collect 90% of the sun's energy.

Arrays can be fixed or moveable. Let's discuss fixed PV arrays first. The simplest location for a fixed array is on the ground. Ground mounting allows you to check out your system easily and adjust the array for the different seasons. However, a ground mount is also an easy target for vandals and thieves, or might attract the curious to tinker with it. Also, lawn mowing or sports activities can lead to an accident. Most arrays have tempered glass or strong plastic glazing, but why take chances?

Some people mount their arrays on poles or windcharger towers to keep them out of reach. If your windcharger tower is flimsy, don't risk your PV array as a strong wind might put you out of business.

The roof is the logical place for a solar collector as it is generally clear of solar obstructions, out of reach, and allows the leads to

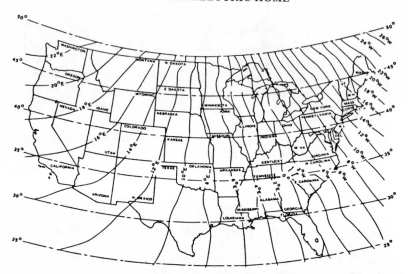

FIGURE 3.4—Map showing the difference between magnetic north and true
  north. (For example, central Texas magnetic north compass reading is
  10° east of true north.)

the battery bank to be short. Be sure to use stand-offs and metal
racks to mount the array. Check with your insurance agent to get
details on home additions, and check with your local building
inspection office for help. You might have to educate building
inspectors as they are not all familiar with solar energy systems.

Tilting your array to get the most solar energy year-round is a
good idea. However, it is not necessary. If your electric usage
is fairly constant year-round, then tilt your array to the same
angle as your latitude. If you use more electricity in winter, as
most of us do because of increased lighting demands, then tilt
the array to latitude plus 15°. Some smaller arrays should be
tilted more often to maximize the amount of solar energy
accepted by the array. If your PV array is easily accessible, you
may want to tilt it four to six times a year. Wait for a sunny day
and at noon adjust your array legs so that the array faces the sun
directly. The sun is highest in the sky on June 21st and lowest in
the sky on December 21st.

Tracking the sun from east to west can increase your electrical
production 20 to 40%, depending on the time of year. For very
small arrays that are easily accessible this may be a good way to
go. You can tilt your array manually simply by turning it toward

the east when you wake up. At 10:00 a.m. turn the array to the south. Around 2:00 p.m. tilt the array to the west. Some people have spring-loaded their PV arrays to face south—by pulling on a string or cable for either east or west, the array is adjusted. Remember, keep it simple.

Automatic tracking is also possible with commercially available sun-actuated motors and clock arrangements similar to the ones used by astronomers. One of the best devices I have seen for east–west tracking is available from Zomeworks. Steve Baer, solar energy pioneer and founder of Zomeworks, has developed a simple freon-driven tracker array rack (Figure 3.5). Freon, a refrigerant with a low boiling point, adjusts the array as it changes from a heavy liquid to a light gas in the two drive chambers of the device.

FIGURE 3.5—An ARCO Solar panel on a Zomeworks passive tracker at the home and workshop of Paul Wilkins.

[Photo: A. D. Paul Wilkins, Sclar Works!]

There is a relationship between your array size, your battery bank size and your array tilt. In general, it is wise to maximize winter collection via tilt and thus cut down on the amount of battery storage required in winter. Be sure to remain flexible in the beginning so that you can adjust your array to its final position.

A practical application . . . .

This home's photovoltaic system supplies an annual average of 1680 watt-hours/day. It powers lights, color TV, radio, vacuum cleaner, toaster, blender, washer, power tools, and water pumped from a 180-ft well. Propane powers a refrigerator, water heater, dryer and stove. Located in Piute Mountains, California.

[Photo: ARCO Solar, Inc., and Greg Johanson, Solar Electric Systems]

CHAPTER 4
SIZING A PV SYSTEM

Sizing a residential PV system is not complicated. In this chapter you will learn how to "think through" the sizing process so that you can adapt the information to suit your specific needs. It must be remembered that this sizing process is for stand-alone or remote home PV installations with battery storage. Grid-connect PV systems or large installations can follow the same basic sizing process, but more variables and different costs will be involved. Our goal is to give you sufficient information to make intelligent choices and decisions as you balance your needs, available hardware, budget and goals.

An important variable is the weather. We will try to average out that local and immediate phenomenon called "weather" so that you can design your PV system for year-round use. This means you will be using "climate" and general climatic data based upon averages. If you have just moved to your homesite, it is wise to talk to as many long-time residents as possible to learn about the climate. You may learn that your valley or ridge is subject to more clouds and fog than the rest of the region—in your situation, an important thing to know. But remember, people tend to exaggerate and forget—so rely on good judgment, too.

The sizing process can be broken down into three main steps:

1. Determine your loads or power requirements.
2. Size your PV array.
3. Size your battery storage.

Each step can be further broken down into simpler steps, and you will be doing just that as you learn the PV sizing process.

If you plan to use other energy inputs (*e.g.*, windcharger, gas generator, hydroelectric generator), it is wise to determine your PV system size first and then examine the other possibilities. That way you can eventually phase out any mechanical or fossil fuel-powered production units.

DETERMINE YOUR POWER REQUIREMENTS

Now we are going to quantify your power requirements. This simply means that you are going to:

1. list the devices you wish to power,
2. list the power requirements for each device, and
3. list the number of hours the device is to be used.

Example 1

Let's begin with a small PV-powered cabin owned by outdoor, nature-loving people who prefer simplicity. The cabin is a passive solar building heated with wood. The occupants use wood or propane for cooking and refrigeration. They have jobs outside the home, and only hand tools are used in and around the residence. The occupants have a collection of recorded tapes. Water is supplied by a gravity feed spring.

Here is a listing of the devices used in the cabin:

> Two livingroom lamps (30-watt fluorescent) used 4 hours per evening
> One kitchen lamp (30-watt fluorescent) used 2 hours per day
> One bedroom reading lamp (0.75-amp incandescent) used 1 hour per night, average
> One automobile am/fm radio tape recorder (0.5-amp) used 4 hours per day, average

The total power requirement for this PV system is:

2 x 1.7 x 4 = 13.6 ampere-hours per day ⎫
1 x 1.7 x 2 = 3.4 ampere-hours per day ⎬   lighting
1 x 0.5 x 1 = 0.5 ampere-hours per day ⎭
1 x 0.5 x 4 = 2 ampere-hours per day (radio/tape player)

Total ampere-hours per day = 19.5.

Example 2

Our second sizing example is a large country home near Phoenix, Arizona, occupied by a family of four. It has three bedrooms, a solar greenhouse, and full indoor plumbing. This time we will make two separate load requirement lists—one for winter and one for summer—as the family's activities differ with the seasons. Spring and fall load requirements are an average of the two quantified seasons.

Lighting
Four each 30-watt fluorescent lamps 6 hours per night
Two each 30-watt fluorescent lamps 2 hours per night
Four each 0.75-amp reading lamps used 2 hours each per night, average

Entertainment
One 12-inch black and white TV used 4 hours per day, average
am/fm radio (0.5-amp) used 4 hours per day
Record turntable (0.5-amp) used with amplifier (0.5-amp) for 2 hours per day, average

Washing machine
Three one-half hour loads three times per week
(3 x 22 amps x 0.5 hour x 3 times per week = 99 ampere-hours per week) totaling (99 ÷ 7 days) 14.2 average ampere-hours per day

Water pump used in pressure system
4.5-amp draw for 1.5 hours per day, average

Digital alarm clock
0.3 ampere-hour per day

120-volt AC loads (powered by a 550-watt inverter):
Sewing machine for 0.5 hour per day
Vacuum cleaner for 0.5 hour per week or 0.13 hour per day
Small power tools used 0.5 hour per day, average
(Assume that inverter loads draw full inverter current to take into allowance time when inverter is drawing standby current.)

Since there are more daylight hours in summer, lighting requirements are one-half winter needs. However, power tools are used three times as much in summer. In addition, two 8-inch fans (1.4-amp) are used 4 hours per day on the average. All other loads remain the same.

The winter power requirement for this PV system is:

4 x 1.7 x 6 = 40.8 ampere-hours per day
2 x 1.7 x 2 = 6.8 ampere-hours per day   lighting
4 x 0.75 x 2 = 6 ampere-hours per day
1 x 1.4 x 4 = 5.6 ampere-hours per day (TV)
1 x 0.5 x 4 = 2 ampere-hours per day (radio/tape player)
1 x (0.5 + 0.5) x 2 = 2 ampere-hours per day (turntable)
14.2 ampere-hours per day (washing machine)
1 x 4.5 x 1.5 = 6.8 ampere-hours per day (water pump)
0.3 ampere-hour per day (digital clock)
(0.5 + 0.13 + 0.5) x 46 = 52 ampere-hours per day
   (inverter load)

Winter total ampere-hours per day = 136.5.

The summer power requirement for this PV system is:

4 x 1.7 x 3 = 20.4 ampere-hours per day
2 x 1.7 x 1 = 3.4 ampere-hours per day   lighting
4 x 0.75 x 1 = 3 ampere-hours per day
9.6 ampere-hours per day (entertainment—same as winter)
14.2 ampere-hours (washing machine—same as winter)
6.8 ampere-hours per day (water pump—same as winter)
0.3-ampere-hour per day (digital clock—same as winter)
52 ampere-hours per day (summer inverter load)
2 x 1.4 x 4 = 11.2 ampere-hours per day (fans)

Summer total ampere-hours per day = 120.9.

As you can see, determining your power requirements is simply a process of listing each load or device, the power it consumes, and the number of hours it will be in use. We have converted watts to amperes to simplify the sizing of the PV array and the battery storage. Just use the Ohm's law chart (Appendix A) or your multi-tester to obtain estimated or actual current draws.

You will also note that we have increased some power loads. In the second example, the actual mathematical answer for the washing machine load is 14.1428 ampere-hours per day. We rounded this off to 14.2 ampere-hours per day to simplify the math. To allow for the possibility of wire resistance loss and variations in usage patterns, it is a good idea to avoid conservative numerical estimates.

In the second example, notice that there is a significant difference (15.6 ampere-hours per day) between winter and summer daily power requirements. Now is a good time to point out that, in general, winter power requirements are higher than those of summer. This is because days are shorter in winter and the extra time spent indoors increases lighting requirements significantly. Of course, this may not be true in your situation—each of us has different energy needs.

These examples are just guidelines to help you list your own actual power requirements. Later on, you may find that you have to adjust your power requirement figures to meet your budget. In that case, it's not a bad idea to make a couple of power requirement lists. You should have one list of your present power requirements—what you are actually using now. Be accurate and be honest. If you are on the utility power grid, list every item in your home and compare (for an accuracy check) your figures with the kilowatt figure on your electric bill.

When you list your PV power requirements, don't shortchange yourself. List everything you want to use. You may discover that you can't afford the array you want, but that's OK. The nice thing about PV systems is that you can start out small and enlarge them as your income allows. And besides, once you do get started, you may find that your power needs are less than your projections. In any case, you will be striking a balance between the ideal and the real. So begin with accurate and honest figures.

SIZE YOUR PV ARRAY

After you have determined your power requirements, the next step is to size your PV array to meet them. The following procedure has been modified from the ARCO Solar design data. The Multiplier Factors Map (Figure 4.1) is a convenient tool for sizing your PV array, and is based on the more detailed solar radiation maps (Figures 4.2 and 4.3). When using the numbers on these maps, be sure to pay attention to the units represented. A multiplier factor is just a number used for convenience to simplify the sizing process. The solar radiation maps list actual and average *measurable* units. Keep this difference straight so that you won't make the mistake of multiplying apples by oranges.

FIGURE 4.1–Multiplier Factors Map, continental United States.
[Courtesy: ARCO Solar, Inc.]

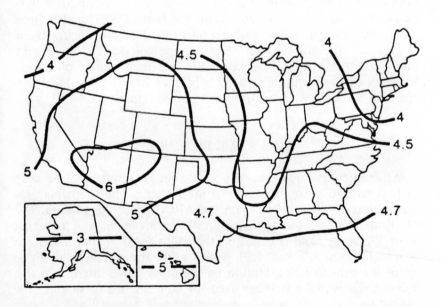

FIGURE 4.2–Peak sun hours per day: yearly average.
[Courtesy: DOE]

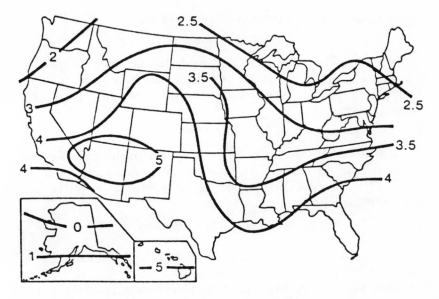

FIGURE 4.3—Peak sun hours per day: 4-week average, 12/7–1/4.
[Courtesy: DOE]

The listing at the end of the chapter gives this information for a number of specific locations worldwide.

Step One. You have already determined your daily power or load requirement. Now you must divide your daily power requirement by 24 (hours) to get the *continuous load amps.*

Step Two. Multiply the continuous load amps by the multiplier factor for your PV array location. This will give you the power requirements for your PV system in *peak amps.*

Step Three. Divide the peak amps by the peak amps of the PV panel you propose to use. Use the manufacturers' peak power current (amperes) rating (@ 100 mW/cm² and 25°C). You now have the number of PV modules required for your 12-volt DC system. If the figure is a fraction, round it off to the next highest whole number. If your PV system is 24, 36 or 48 volts, then connect your 12-volt modules in series to get the proper voltage. A 24-volt system will require twice as many modules as a 12-volt system with the same peak ampere requirement.

Now let's see how this simplified method works. We will use Example 1—the small cabin that requires 19.5 total ampere-hours per day.

Step One. Compute the continuous load amps.

$$19.5 \div 24 = 0.8125 \text{ (round off to } 0.82)$$

Step Two. The cabin is located near Omaha, Nebraska. We can see that the multiplier factor for this location is 7. Multiply the continuous load amps by the multiplier factor to get the peak amps requirement.

$$0.82 \times 7 = 5.75$$

Step Three. Divide the peak amps by the manufacturer's peak ampere rating for the PV modules to be used. In this example, we will use the ARCO Solar ASI 16-2000 rated at 35 watts, 2.26 peak amps.

$$5.75 \div 2.26 = 2.54 \text{ PV modules}$$

Thus, this system will require three ASI 16-2000 PV modules. By rounding off the number of modules required to the next whole number, you ensure that the system will have adequate power during winter, with a surplus of power in summer.

If your budget does not permit the purchase of three modules at this time, you can squeeze by with two and a combination of careful battery monitoring and conservation. Another way to stretch the power from a limited PV array is to track the sun. Manual east-to-west tracking or the use of an automatic tracker, such as the freon-driven solar tracker manufactured by Zomeworks, will increase the PV array output approximately 15% in winter and up to 40% the rest of the year at the 35°N latitude.

Now let's compute the size of the PV array for the larger home in Example 2. To ensure that the home has adequate power year-round, we must use the winter total ampere-hours per day (89.7). Using our formula and the Multiplier Factors Map, all we need to know is the peak ampere rating of the PV module to be used and the location of the home. In this case we will be using the Photowatt ML7010 55-watt PV module rated at 3.35 amps current at peak power.

Step One. $89.7 \div 24 = 3.8$ continuous load amps

Step Two. 3.8×5 (multiplier factor) = 19 peak amps

Step Three. $19 \div 3.35 = 5.67$ for a total of six PV modules

SIZE YOUR BATTERY STORAGE

When attempting to design the optimum PV system, you will find that there is a trade-off between the number of PV modules in your array and the size of the battery storage. There is also a relationship between array tilt and storage size. In general, a fixed array is tilted to boost winter output at the expense of summer output. This levels out the system's performance over the year and decreases the amount of battery storage needed to get through winter months. Such a design is optimum for steady, year-round power. If power needs vary significantly with the seasons, then the storage, tilt and array size must be designed to accommodate such variations.

A fixed PV array will be tilted just like a fixed solar water heater—that is, latitude plus 15°. But sizing the battery storage is a complex trade-off based on a few simple principles. If there were no clouds, no seasons, and no limits to battery performance, you would need only enough storage to last through the night. But in the real world, you will need more.

First, there will be cloudy days when the solar array will not be able to fully recharge the batteries. You must have enough capacity to last through several overcast days in a row. As discussed elsewhere, five days of autonomous storage is the minimum in the continental United States, except for the sunny southwest. Second, there are fewer hours of usable sunlight in winter and a battery's average state of charge will be lower during winter. Even at its predicted lowest, your storage must have at least as many ampere-hours of "required" storage as it does autonomous storage. That means, if you need 200 ampere-hours of storage to take you through cloudy periods, you will need at least 200 ampere-hours additional required or reserve battery capacity.

Third, battery performance has its limits. Repeated cycling between charging and discharging will eventually shorten battery life. Both sulfation (growth of lead sulfate crystals on the battery's plates) and stress fatigue can be partially prevented by adding more capacity so that the depth of discharge never gets below 60% in winter. However, your batteries will get old and weaker over the years. They will last longer and work better if protected from extremes of heat and cold—especially cold. People locate their batteries in the house, in the enclosed crawlspace, and even in a well drained underground chamber to take advantage of everything from the earth's thermal inertia to normal house heat. If you don't do this, you must increase your battery capacity.

All of these factors must be considered—they add up to a lot more than one night's storage. In the very sunniest locations, the battery capacity will be 7 to 10 times larger than the average daily load. The stormier climates can require 20 to 30 times the daily load. If your area has a history of weeks of cloudy and partly cloudy weather, seriously consider the trade-off between the cost of additional battery storage (twice as much as you might need in some areas) as compared to the regular cost of battery replacement every three to five years. Some battery banks have lasted a decade and longer—and they were used telephone company batteries to begin with.

If your batteries will experience freezing temperatures regardless of preventive measures, the battery storage capacity should be increased. Table 4.1 will be helpful in allowing for freezing temperatures experienced over extended periods of time. Find the lowest extended time period temperature for your area and multiply your calculated storage capacity by the correction factor.

Table 4.1—Temperature Correction Factors

Temperature °C	Correction Factor	Temperature °F
−10	1.10	+14
−15	1.55	+ 5
−20	2.05	− 4
−25	2.75	−13
−30	3.50	−22
−35	4.25	−31

There are many methods for calculating the amount of battery storage you will need. A basic rule of thumb for approximately three days battery storage is to size the battery bank based on the daily total ampere-hours per day requirement. The ratio is 10 to 1. For every 1 ampere-hour daily load requirement, you will need 10 ampere-hours of battery storage. The example cabin requires 19.5 ampere-hours per day times 10 for a total of 195 ampere-hours of battery storage.

Let's see how useful this rule of thumb is. In winter, the cabin's location can have up to seven cloudy days in a row. That means we need at least 136.5 usable ampere-hours of storage in the battery bank. If we want our battery bank to last a long time, we must not deep-discharge the batteries too often or too deeply. It is recommended that you use no more than 40% of your battery storage capacity during discharging or cycling. Further, you will experience a 30% loss of battery-rated storage due to internal battery resistance, self-discharge and wire resistances in your system. Thus, the size of the battery storage needed for the example cabin should be at least 325 ampere-hours. As 6-volt deep-cycle (golf-cart) batteries usually come in 200-ampere-hour size, this system would need four such batteries. The battery configuration would be two sets of parallel batteries connected in series.

APPLIANCES, DEVICES AND LOADS

The words appliances, devices and loads are interchangeable as we are using them here. This section will describe some of the hundreds of loads possible with a PV power system. Just a look at the index of any J. C. Whitney catalog will reveal the range of loads possible.

All of the items below are 12-volt DC. If inverters and low-wattage standard 120-volt AC equipment were included, this list would be much longer.

air compressors	electric door locks	polishers/sanders
air conditioners	electric oil changers	radios
alarm timers	electric window lifts	refrigerators
amplifiers	fans	relays
battery analyzers	flashers	revolving lights
battery testers	float, tank	roof fans
blowers	fluorescent lamps	sirens
bug killers	hair curlers	spark plug cleaners
burglar alarms	hair driers	spotlights
CB equipment	heaters	stereo tape players
cigarette lighters	hoists	stereo equalizers
circuit breakers	horns	stereo amplifiers
clocks	intercoms	trouble lights
coffee makers	lights, all types	vacuum cleaners
defroster blowers	lights, quartz	water heaters
digital clocks	lights-on alarms	water pumps
door jamb switches	motors	winches

As you can see, there are dozens of things you can use with your 12-volt DC system, just from one supplier. When you start to count the different styles, sizes and manufacturers, there are literally hundreds.

Listed below is the power consumption of some of the more common 12-volt DC loads. Use this list when determining your power requirements and sizing your system.

Lights

Fluorescent 12-volt DC
    16-watt (two 8-watt bulbs) 1.4 amps
    30-watt (two 15-watt bulbs) 1.9 amps
    22-watt circular 1.4 amps
Incandescent 12-volt DC type
    15-watt standard base 1.3 amps
    50-watt standard base 4.2 amps
    (screw base and automotive type rated on bulb or package)
Quartz-iodine, metal halide or tungsten-halogen
    17-watt 1.5 amps
    34-watt 2.9 amps

Entertainment/Communications

12-inch black and white television (16-watt) 1.4 amps
13-inch color television 4.1 amps
am/fm car radio with tape player 0.5 amp
Turntable (converted to 12-volt DC) 0.5 amp
Radio/telephone
    receive 0.3 amp
    transmit 2.5 to 15 amps
CB radio
    receive 0.3 amp
    transmit 0.5 amp
Intercom 0.5 amp

Appliances

Coffee pot 11 amps
Popcorn popper 16 amps
Toaster 20 amps
(Mixer, blender, can opener, food processor, etc.—see inverter section in in this chapter)
Washing machine (wringer-style with 12-volt DC ¼-hp motor) 22 amps
Refrigerator
    12-volt DC Koolatron (3 cubic feet) 2 amps
    Solarwest WSP-12-6 (3.7 cubic feet) 1 amp
    Solarwest 1E-12V (2.8 cubic feet) 2.1 amps
    Danfoss compressor BD2.5 (up to 8 cubic feet) 3.7 to 6.3 amps
    Norcold model 8010 (9 cubic feet) 20 amps
Vacuum cleaner
    Sears 12-volt DC (108-watt) 9 amps
    J. C. Whitney hand-type 4.5 amps
    J. C. Whitney upright 20 amps

Fans
    8-inch blade, oscillating 1.4 amps
    6-inch vent fan 2 amps
    Converted 12-inch fan with Datsun three-speed blower motor 4.5 amps
    (motor available from Airborne Sales)

Tools

    Chain saw (Minibrute) 100 amps
    Drill, 3/8-inch 12 amps
    Air compressor (60-watt) 5 amps
    Polisher/sander 7 amps
    Hoists up to 100 amps
    Cordless tools with ni-cad power packs, as rated
    Motors, as rated

Water Pumps

    March 893 0.9 amp
    Honeywell ¼-hp with separate pump 11 amps
    Jabsco (4 gallons per minute) 4.5 amps
    J. C. Whitney nonautomatic (2 gallons per minute) 1.5 amps

Miscellaneous

    Water purifier (24-watt) 2 amps
    Bug killer (12-volt DC) 1.2 amps
    Travel iron 10 amps
    Curling iron (40-watt) 3.5 amps
    Digital alarm clock 0.3-ampere hours per day total (continuous duty)

INVERTER LOADS

There is a temptation to try to use all 12-volt DC equipment or to try to use conventional 120-volt AC equipment exclusively. Both are possible with a PV power system. However, in the first case you will find that not all equipment comes in 12-volt DC, and trying to convert everything to 12-volt DC can be difficult even for the most handy among us. At the opposite extreme is the PV system with a large inverter or several small inverters for powering conventional appliances and equipment. This method, generally promoted by the DOE photovoltaics projects, can be expensive and wasteful.

Just as your choice of appliances and equipment is based on personal needs, so too should the power you select fit your needs. In almost all cases it is less expensive to use 12-volt DC fluorescent

FIGURE 4.4—Sewing machine and blender modified to 12-volt DC.
[Photo: A. D. Paul Wilkins, Solar Works!]

and quartz lights rather than incandescent bulbs. The use of 12-volt DC water pumps with a cistern water supply is more efficient than other methods. But trying to make or convert some devices to 12-volt DC can be more costly than using existing 120-volt AC equipment and an efficient solid-state inverter.

A 550-watt Tripp-Lite inverter allows you to use equipment from your pre-PV days. You can also use low-cost new or second-hand appliances that operate on 120-volt AC. Small power drills and saber saws work well. A zigzag sewing machine motor draws only 0.85 amp and gives much more versatility than a treadle machine. Many kitchen devices which are used intermittently, such as blenders, mixers, can openers and electric knives, can be powered by an inverter.

When selecting an inverter for the occasional 120-volt AC load, be sure to consider what you will be powering. AC motors often draw five times the running or rated operating current to start, so the inverter must be sized to carry that initial load. If you have a 120-volt AC grain mill, you might consider replacing the motor with a DC motor, as the original AC motor is probably an induction-type and will need a big inverter to power the start-up.

Try to match your devices to the inverter. Upon examination of the power requirements of a collection of 120-volt AC devices, most of the smaller ones were found to be in the 2- to 3-amp range. It was less costly to replace the larger motors—such as those for pumps and large power tools—than to buy the big inverter they would require to power them.

FIGURE 4.5—12-volt DC deep-well water pump designed and built by Paul Wilkins.
[Photo: A. D. Paul Wilkins, Solar Works!]

TYPICAL LOADS

The following lists are examples of actual homes and the equipment used in them. The equipment in your home may be similar to one of these examples. Remember, either list the devices you presently use or intend to use. This is not a "dream home" we're talking about, but your actual home and your real needs.

Example 1

ampere-hours/day

Two 16-watt fluorescent lamps used 4 hours daily	1.2 x 2 x 4 = 9.6
Automotive am/fm/cassette used 4 hours daily	0.5 x 4 = 2
8-inch oscillating fan used 4 hours daily	1.4 x 4 = 5.6
	Total = 17.2

Example 2

	ampere-hours/day
Three 16-watt fluorescent lamps used 4 hours daily	1.2 x 3 x 4 = 14.4
Two 30-watt fluorescent lamps used 4 hours daily	1.7 x 2 x 4 = 13.6
Automotive am/fm/cassette used 4 hours daily	0.5 x 4 = 2
Turntable used 2 hours daily with radio as amplifier	0.5 + 0.5 x 2 = 2
12-inch black and white TV used 6 hours daily	1.4 x 6 = 8.4
Inverter (550-watt, intermittent use)	
sewing machine 0.5 hour daily average	
(3-amp inverter draw)	3 x 0.5 = 1.5
vacuum cleaner 0.5 hour weekly average	
(4-amp inverter draw)	4 x 0.5 ÷ 7 = 0.3
blender 0.5 hour weekly average	
(3-amp inverter draw)	3 x 0.5 ÷ 7 = 0.2
Water pump used 1 hour daily	4.5 x 1 = 4.5
	Total = 46.9

DOMESTIC PHOTOVOLTAIC POWER

This is another example of a solar electric home design. The design option section at the end of this section is a good example of the "cut and try" reasoning a person should go through to arrive at a properly balanced system.

[From "Solar Photovoltaic Application Seminar: Design Installation and Operation of Small, Stand-Alone Photovoltaic Power Systems," July 1980.]

General

The PV system will provide electricity for remotely located homes where hook-up to utility grid lines is not economically feasible.

Refrigeration

A 6-cubic foot refrigerator will provide cold storage for the average family; 12-volt DC motor, 60 watts, 5 amps, 25% duty cycle average. Total ampere-hours required: 24 hr x duty cycle x 5 amps = 30 ampere-hours.

Lighting

The home is equipped with four fluorescent lamps—two 40-watt and two 10-watt—used 4 hours per night (a conservative estimate). 2 x 40 watts ÷ 12 volts x 4 hr + 2 x 10 watts ÷ 12 volts x 4 hr = 33.3 ampere-hours.

Television

Television usage is estimated at approximately 4 hours per day at 24 watts equaling 8 ampere-hours (4 hr x 24 watts ÷ 12 volts = 8 ampere-hours).

Cistern Water Pump

A permanent magnet positive displacement water pump is used, which draws 5 amps under full load. The pump fills a pressurized tank, which then supplies the home's water needs. A positive displacement pump was chosen over a centrifugal or screw-type pump due to its higher efficiency (80% versus 35 to 50%). The family's water requirements are 200 gal/day; the pump provides 9.4 gal/min, requiring 6.38 ampere-hours (200 gal/day is pretty high and the flowrate should be reduced to cut water usage.)

Total Charge Requirements Per Day

77.7 ampere-hours per day

System Component Sizing

Array—220 peak watts or 13.75 peak amps
Battery—680 ampere-hours at 12 volts DC
Voltage regulator to prevent excessive outgassing due to overcharging

Notes

1. A manual switch isolates the load for maintenance purposes.
2. The system is fused to prevent battery drain if a short circuit develops. Standard buss fuses (car fuses) are acceptable and readily available.
3. A voltage meter with its switches located in the battery storage box isolates and monitors the battery or array voltage.
4. An ammeter with switches located in the battery storage box monitors total array output current or total load current.
5. Electrical storage is provided by low self-discharge lead-acid batteries.
6. Venting or recombiner caps for the batteries prevent dangerous hydrogen gas build-up.
7. Batteries may undergo a maximum 60% depth of discharge.
8. Disconnect switches are not required between the PV panels in the array due to low system voltage (12-volt DC).
9. An opaque cover should be used during maintenance to cover and effectively "turn off" the array.
10. Due to the location of the battery storage in the home, special care must be taken to ensure proper venting of gases and protection of the battery terminals against accidental shorting.

Design Options

1. Using 11 each PV panels (12-volt DC, 20 peak watts @ $11.25/peak watt = $225) seasonal storage is not needed, tilt angle used is latitude plus 15°.
2. Using 10 panels, seasonal battery storage needed is 128.5 ampere-hours at latitude plus 15° tilt.
3. Tilting 10 panels at the degrees latitude increases seasonal storage to 402.2 ampere-hours.
4. Optimal choice: between one extra panel or 128.5 additional ampere-hours of battery storage. (Battery cost is $170/kWh @ 500 hour rate @ 60% depth of discharge = $437).
5. Difference is $437 – $225 = $212.
6. Percent difference is 212/225 = 94.27%.
7. Conclusion: choose the additional panel.

35-Watt PV Module Operation by Country

Lat.	Long.	Location	Ampere-Hours Generated		Tilt Angle
			Per Day	Per Week	
27N	1E	Algeria Aoulef	13.4	93.7	45S
30N	2W	Algeria, Beni-Abbes	13.1	91.5	45S
34N	1E	Algeria, Chottech Cherqui	12.2	85.6	45S
25N	1E	Algeria, Quallen	14.5	101.4	30S
23N	6E	Algeria, Tamanrasset	14.0	98.1	30S
30N	31E	Arab Rep. of Egypt, Giza	12.5	87.3	45S
41S	71W	Argentina, Bariloche	8.7	61.1	65N
35S	58W	Argentina, Buenos Aires	10.2	71.3	60N
28S	59W	Argentina, Corrientes	10.7	75.2	50N
31S	62W	Argentina, Rafaela	7.1	50.0	55N
32S	69W	Argentina, San Juan	11.2	78.5	45N
35S	139E	Australia, Adelaide	11.0	77.2	60N
35S	149E	Australia, Canberra	11.1	77.8	60N
38S	145E	Australia, Melbourne	9.6	67.3	60N
32S	116E	Australia, Perth	10.9	76.2	60N
34S	152E	Australia, Sydney	11.9	83.5	45N
19S	147E	Australia, Townsville	12.5	87.5	30N
48N	16E	Austria, Vienna	6.4	45.0	70S
17S	68W	Bolivia, La Paz	12.6	88.0	30N
20S	23E	Botswana, Maun	12.7	88.7	20N
43N	28E	Bulgaria, Varna	7.9	55.2	45S
59N	94W	Canada, Churchill	9.0	63.2	80S
45N	64W	Canada, Dartmouth	8.1	56.4	65S
54N	114W	Canada, Edmonton	9.2	64.2	75S
62N	121W	Canada, Fort Simpson	7.5	52.4	85S
46N	74W	Canada, Montreal	7.8	54.8	70S
44N	79W	Canada, Toronto	8.1	56.4	65S
49N	132W	Canada, Vancouver	6.8	47.4	70S
12N	15W	Chad, Fort Lamy	14.1	98.9	15S
23S	69W	Chile, Atacama	15.7	109.9	35N
34S	71W	Chile, Santiago	10.6	74.1	55N
44N	125E	China, Changchun	9.4	65.6	65S
38N	122E	China, Chefoo	9.8	68.4	55S
31N	121E	China, Shanghai	9.5	66.3	45S
5N	74W	Colombia, Bogota	10.0	69.8	15S
4S	15E	Congo Rep., Brazzaville	9.8	68.6	15N
14N	89E	El Salvador, San Salvador	13.2	92.7	15S
63N	26E	Finland, Luonetjarvi	5.9	41.5	85S
43N	5E	France, Marseilles	9.4	65.8	65S
49N	3E	France, Paris	6.9	48.4	70S
52N	7E	Germany, Bochum	4.8	33.7	70S
48N	12E	Germany, Munich	7.7	53.8	70S
6N	0E	Ghana, Accra	10.6	73.9	15S
38N	24E	Greece, Athens	9.7	68.2	55S

Lat.	Long.	Location	Ampere-Hours Generated		Tilt Angle
			Per Day	Per Week	
12N	16W	Guinea-Bissau	12.7	88.7	15S
22N	114E	Hong Kong	9.8	68.6	15S
19N	73E	India, Bombay	13.0	91.1	20S
23N	88E	India, Calcutta	11.6	81.1	25S
13N	80E	India, Madras	13.4	93.7	15S
6S	107E	Indonesia, Djakarta	9.8	68.6	15N
32N	46E	Iraq, Al-Kut	12.4	87.0	55S
32N	35E	Israel, Jerusalem	12.5	87.8	55S
41N	14E	Italy, Napoli	8.1	56.9	65S
46N	9E	Italy, Pallanza	8.8	61.3	65S
35N	137E	Japan, Nagoya	9.9	69.4	30S
43N	141E	Japan, Sapporo	7.6	53.3	65S
36N	140E	Japan, Tokyo	7.2	50.2	45S
1S	37E	Kenya, Nairobi	11.4	79.5	14N
35N	129E	Korea, Puzan	11.5	80.4	30S
38N	127E	Korea, Seoul	9.6	67.0	55S
34N	36E	Lebanon, Ksara	12.2	85.2	45S
15N	146E	Mariana Islands, Saipan	12.2	85.2	15S
17N	7W	Mauritania, Nema	12.2	85.7	20S
20N	99W	Mexico, Mexico City	11.8	82.4	15S
28N	107W	Mexico, Nonoava	13.7	96.0	35S
23N	110W	Mexico, Cabo San Lucas	12.1	84.5	30S
20N	106W	Mexico, Tomatlan	11.7	81.9	30N
18N	93W	Mexico, Tuxtla Gutierrez	9.6	67.0	15S
17N	100W	Mexico, Acapulco	12.1	85.0	15S
28N	13W	Morocco, Cabo Judy	11.1	77.6	35S
23S	17E	Namibia (SWA), Windhoek	14.5	101.2	30N
30N	85E	Nepal, Saga	11.4	79.5	30S
4S	139E	New Guinea, Baliem	12.8	89.4	15N
9S	140E	New Guinea, Merauke	10.4	72.6	15N
4S	152E	New Guinea, Rabaul	11.1	77.9	15N
7S	147E	New Guinea, Bulolo	8.5	59.5	15N
18S	177E	New Zealand, Nandi	12.6	88.5	30N
29S	178E	New Zealand, Raoul Is.	10.7	74.8	50N
41S	175E	New Zealand, Wellington	8.7	61.1	65N
17N	8E	Niger, Agadez	15.1	105.4	15S
7N	6E	Nigeria, Genin City	8.8	61.3	15S
9N	12E	Nigeria, Yola	12.2	85.6	15S
60N	5E	Norway, Bergen	5.2	36.4	75S
25N	67E	Pakistan, Karachi	11.8	82.6	25S
30N	67E	Pakistan, Queta	13.2	92.3	50S
9N	80W	Panama, Panama	10.5	73.2	15S
12S	75W	Peru, Huancayo	14.4	100.7	15N
15N	121E	Philippines, Quezon City	9.7	67.7	15S

Lat.	Long.	Location	Ampere-Hours Generated		Tilt Angle
			Per Day	Per Week	
38N	9W	Portugal, Lisbon	11.3	78.8	60S
19N	66W	Puerto Rico, San Juan	13.6	95.5	20S
26N	50E	Saudi Arabia, Dhahran	12.6	88.3	45S
24N	50E	Saudi Arabia, Riyadh	13.2	92.5	30S
15N	17W	Senegal, Dakar	11.6	81.1	15S
9N	13W	Sierra Leone, Longo	10.4	72.7	15S
1N	104E	Singapore	9.2	64.2	15S
41N	4W	Spain, Madrid	9.6	67.2	65S
14N	25E	Sudan, El-Fasher	12.7	88.9	15S
20N	37E	Sudan, Port Sudan	12.7	89.0	30S
63N	14E	Sweden, Froson	6.9	48.4	75S
56N	13E	Sweden, Svalov	6.1	42.9	75S
25N	122E	Taiwan, Taipei	7.9	55.5	30S
23N	120E	Taiwan, Tainan	11.1	77.6	15S
24N	122E	Taiwan, Kwarenko	10.1	70.6	25S
14N	101E	Thailand, Bangkok	11.1	77.6	15S
3N	35E	Uganda, Moroto	14.2	99.1	15S
29S	17E	Republic, So. Africa, Alexander Bay	13.7	95.8	45N
30S	31E	Republic, So. Africa, Burban	10.7	74.8	35N
34S	18E	Republic, So. Africa, Capetown	12.1	85.0	45N
24S	29E	Republic, So. Africa, Pietersburg	13.2	92.3	30N
26S	28E	Republic, So. Africa, Pretoria	12.6	88.5	30N
29S	21E	Republic, So. Africa, Upington	13.3	93.0	40N
52N	0E	United Kingdom, Cambridge	6.7	42.9	70S
		United States of America			
32N	86W	AL, Montgomery	10.5	73.4	45S
61N	162W	AK, Bethel	8.1	56.9	75S
65N	148W	AK, Fairbanks	8.5	59.5	75S
62N	149W	AK, Matanuska	7.6	52.9	75S
33N	112W	AZ, Phoenix	14.0	98.2	45S
35N	92W	AR, Little Rock	10.0	69.9	55S
39N	122W	CA, Davis	10.9	76.2	60S
37N	120W	CA, Fresno	11.9	83.5	45S
36N	118W	CA, Inyokern	15.4	107.8	45S
34N	118W	CA, Los Angeles	12.2	85.2	45S
34N	117W	CA, Riverside	12.7	89.0	45S
40N	105W	CO, Boulder	10.4	72.7	55S
40N	106W	CO, Grandby	12.1	84.9	55S
39N	74W	DC, Washington	9.0	63.2	60S
30N	82W	FL, Gainesville	11.2	78.6	45S
26N	80W	FL, Miami	12.0	83.8	25S
34N	84W	GA, Atlanta	10.5	73.6	45S
21N	158W	HI, Honolulu	12.8	89.9	35S

Lat.	Long.	Location	Ampere-Hours Generated		Tilt Angle
			Per Day	Per Week	
44N	116W	ID, Boise	10.5	73.4	65S
42N	87W	IL, Chicago	6.7	46.9	65S
40N	86W	IN, Indianapolis	9.0	62.8	60S
42N	94W	IA, Ames	9.4	65.6	65S
38N	100W	KN, Dodge City	12.3	86.4	55S
39N	97W	KN, Manhattan	9.7	68.2	60S
38N	85W	KY, Lexington	10.5	73.8	65S
30N	90W	LA, New Orleans	9.1	63.4	45S
32N	93W	LA, Shreveport	9.9	69.1	55S
44N	70W	ME, Portland	9.6	67.3	65S
42N	71W	MA, Boston	8.2	57.3	65S
43N	84W	MI, East Lansing	8.5	59.7	65S
46N	94W	MN, St. Cloud	9.9	69.4	65S
39N	90W	MO, St. Louis	9.3	65.4	60S
47N	111W	MT, Great Falls	10.5	73.6	65S
41N	96W	NB, N. Omaha	10.2	71.3	65S
36N	115W	NV, Las Vegas	14.2	99.3	45S
40N	75W	NJ, Sea Brook	9.0	62.8	60S
35N	107W	NM, Albuquerque	14.4	101.0	45S
41N	74W	NY, New York City	8.5	59.4	65S
39N	115W	NC, Ely	12.7	89.2	60S
36N	80W	NC, Greensboro	10.0	70.3	55S
47N	101W	ND, Bismarck	10.7	74.8	65S
41N	62W	OH, Cleveland	8.4	58.8	65S
35N	98W	OK, Oklahoma City	11.9	83.5	45S
42N	123W	OR, Medford	9.6	67.3	65S
40N	80W	PA, Pittsburgh	7.0	48.9	65S
41N	71W	RI, Newport	9.0	63.2	65S
33N	80W	SC, Charleston	10.8	75.5	45S
44N	102W	SD, Rapid City	11.2	78.1	65S
36N	87W	TN, Nashville	9.9	69.1	45S
32N	101W	TX, Big Spring	11.8	82.8	45S
33N	97W	TX, Fort Worth	11.8	82.6	45S
41N	112W	UT, Salt Lake City	10.7	74.6	65S
38N	77W	VA, Richmond	8.8	61.6	60S
47N	122W	WA, Seattle	7.9	55.0	65S
48N	118W	WA, Spokane	9.5	66.8	70S
38N	82W	WV, Charleston	7.8	54.5	60S
43N	89W	WI, Madison	9.1	64.0	65S
43N	109W	WY, Lander	12.6	88.5	65S
46N	31E	USSR, Odessa	8.2	57.3	65N
43N	132E	USSR, Vladivostok	10.0	69.8	45S
60N	31E	USSR, Leningrad	6.0	42.3	75S

Lat.	Long.	Location	Ampere-Hours Generated		Tilt Angle
			Per Day	Per Week	
35S	56W	Uruguay, Montevideo	11.0	76.9	60N
10N	65W	Venezuela, Barcelona	11.9	83.5	15S
11N	72W	Venezuela, Maracaibo	11.6	82.6	15S
13N	45E	Yemen, Aden	13.0	91.1	15S
45N	21E	Yugoslavia, Beogard	8.2	57.5	65S
12S	28E	Zaire, Lubumbash	12.1	85.0	15N
1N	25E	Zaire, Kisangani	9.9	69.6	15S
20S	29E	Zimbabwe, Bulawayo	13.0	91.3	20N

CHAPTER 5
SOLAR CELLS & ARRAYS

Solar cells work basically like batteries. A battery gets its energy in a factory or when recharged in a car or truck. A solar cell gets its energy from the sun.

Solar cells produce about one-half volt DC. The amount of current is dependent on the size of the solar cell: larger solar cells produce more current. At present, most cells manufactured are between 0.125 amp and the 2 amps produced by 4-inch cells.

For the practical use of solar electricity, more power is needed. Solar cells are connected in series to increase the voltage. To increase the current, they are connected in parallel. "Load" represents the equipment being powered.

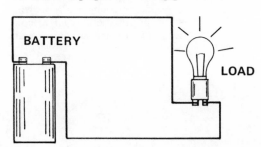

FIGURE 5.1a—A battery converts chemical energy into electricity.

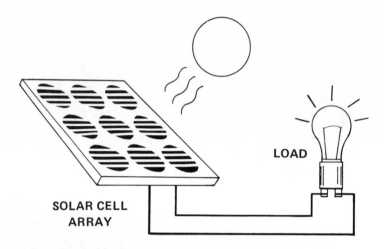

FIGURE 5.1b–A solar array converts light into electricity; nothing internal gets used up or wears out.

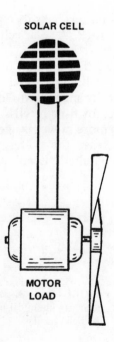

FIGURE 5.2–Small toy motors, like the one shown here, will operate on the output of only one solar cell. This simple device consists of a 2- to 4-inch solar cell, a short length of two wire leads for positive and negative connections, and a small DC motor with blades.

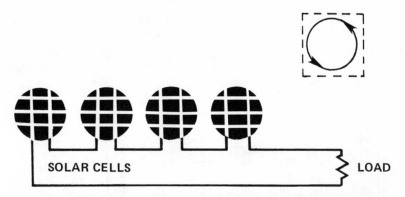

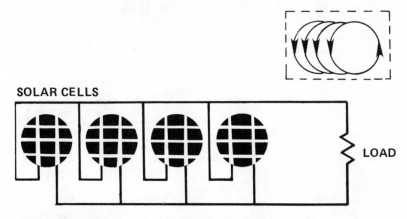

FIGURE 5.3a—Solar cells in series. The total voltage is the sum of the individual cell voltages, but the current is the same as that of a single cell.

FIGURE 5.3b—Solar cells in parallel. Here the voltage would be that of only one cell, but the current is the sum of the individual cell currents.

In a typical 12-volt solar module, 30 to 36 cells are connected in series to produce enough voltage to charge a 12-volt battery. The total voltage of the solar module must exceed the battery voltage to "push" the charge into the battery. Most solar panels produce 14 to 16 volts for battery charging. The battery can then store electricity for periods of cloudiness or darkness. For example, ARCO Solar modules have 33 solar cells each which produce 16 volts at 2 amps open circuit. By connecting modules in parallel to create a PV array, the amount of current is increased. Three ARCO Solar modules in parallel will produce 6 amps.

Solar modules require blocking diodes (see Figure 5.4). A diode is an electrical one-way gate which permits current to flow in one direction but prevents it from flowing in the opposite direction. The diode prevents the battery from being discharged backwards through the solar panel at night.

SOLAR CELL MODULE

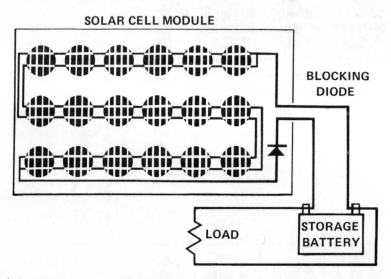

FIGURE 5.4—A solar module, constructed as a single panel with a built-in blocking diode, is shown here connected to both a load and a battery.

When several solar panels are connected in a group you have a solar array. Figure 5.5 shows three panels or modules connected in parallel to increase the current. As shown in this figure, a voltage regulator has been added between the battery and the solar array to prevent overcharging and to protect the battery. Any number of solar cells and batteries can be connected in both series and parallel to produce the needed voltage and current. Care must be taken to match solar cells and batteries because bad cells or low-power cells in both batteries and PV modules will pull down the entire output.

To monitor the PV system, voltmeters and ammeters are installed (see Figure 5.6). Voltmeters are connected parallel to the load. Ammeters are connected in series with the load. A fuse is added in the circuit to protect the load or equipment from damage from a current surge.

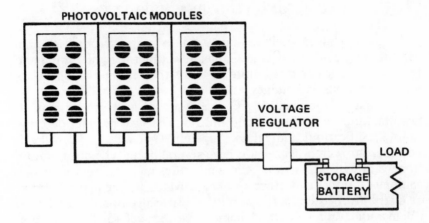

FIGURE 5.5—An array of three solar modules connected in parallel. Usually, each module is a single panel, but some manufacturers make large panels of more than one module. These modules are independent and can be connected in series or parallel, depending on the need.

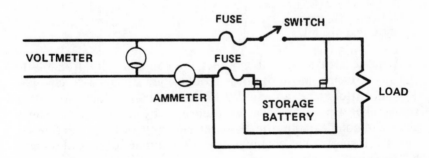

FIGURE 5.6—A voltmeter and an ammeter can be connected to a small solar electric system to monitor the performance of the array and the battery. Note fuse protection and disconnect switch.

ACQUIRING YOUR PV EQUIPMENT

A broad range of modules and associated power production and conditioning equipment is available. Selecting the right equipment to serve your needs is no more difficult than finding a vehicle to meet your needs and budget.

The first step in acquiring your equipment is research: contact manufacturers, distributors and individuals to get product descriptions, specifications and prices. Appendix G and Table 5.1 will aid you in your search. Once you have several packets of information from various sources, note the similarities and differences in the equipment, prices and other services provided by the manufacturer or distributor. Examine the warranties especially closely. A reliable PV manufacturer will stand behind the product for five or more years. Some manufacturers will help you design your system at no charge. It is important to note that their sizing will be on the "heavy" or expensive side for they want to be sure that you will be able to meet your power needs and, of course, buy alot of hardware.

Price is a major factor in buying PV modules. The price of the module is affected by factors such as availability, the quantity ordered, and the desire on the part of the supplier to get equipment in use. Don't forget to add the shipping charges. Finally, consider what other hardware will be needed, such as a regulator, fusing, diodes, and mounting accessories. It all adds up.

Commercial Modules

A number of solar cell modules are commercially available to the homeowner. Most of these modules are designed to charge a 12-volt battery and are similar in their output voltage characteristics. The current and wattage delivered by the module under a given sun condition are the main determinants of its size and price. Table 5.1 gives some characteristics of a number of modules made by different manufacturers. The list is not intended to be all-inclusive but illustrates the variety available. No prices are given because they can change so quickly. You will notice that 35 watts or so is a common size. This is the output of arrays made from the standard 4-inch round cell.

Because photovoltaic systems are inherently modular, you can start with a small system and add modules as your pocketbook allows and needs grow. You can even mix modules from different manufacturers in the same array, but it is best to pick systems with the same design voltage. (This usually means modules with the same number of cells, although some manufacturers' cells produce a slightly higher voltage than others, and some modules require one or two fewer cells to do the same job.) It is a good idea to put a separate blocking diode on each different group of modules.

FIGURE 5.7—Applied Solar Energy provides a wide assortment of One-Sun Solar Electric Modules for remote power requirements.
[Photo: Applied Solar Energy Corp.]

Table 5.1—Commercial PV Modules
(A Comparison of Electrical and Physical Characteristics)*

Manufacturer	Model Number	Peak Output Power (AM1)	Dimensions		Warranty	Remarks
Applied Solar Energy Corporation	60-3062-12	16 volts; 5.2 amps 80 watts	47.24" x 27.38" x 1.54"	3" round cells		Cell wiring: 34 series with four parallel circuits—136 cells
	60-3039-12	18 volts; 1.23 amps 22.2 watts	30.0" x 12" x 1.75"	3" round cells		36 cells
ARCO Solar Inc.	ASI 16-2000	15.5 volts; 2.26 amps 35 watts	48" x 12" x 1.5"	4" round cells	5 years	33 cells
	M51	17.3 volts; 2.31 amps 40 watts	48" x 12" x 1.5"	4" round cells	5 years	35 cells
Chronar Corporation	CPV-4031	22 volts; 2.75 amps 27 watts	36" x 36" x 0.75"	½" square cells	1 year factory	3 each 1' x 3' modules; thin-film substrates with frame
Photowatt International	ML5010	16.5 volts; 2.2 amps 35 watts	48" x 12.5" x 1.5"	4" round cells	2 year limited	35 cells
	ML7010	16.5 volts; 3.35 amps 55 watts	49.5" x 19.5" x 1.5"	5" round cells	5 years	36 cells

Silonex		16 volts; 1.15 amps 18 watts	21.2" x 17.6" x 1.74"	3" round cells		36 cells
Solarex	4330EG	16.7 volts; 2.1 amps 33.5 watts	43.5" x 17.75" x 0.75"	4" round cells	5 years	38 cells
	5330EG	17.5 volts; 2.1 amps 35 watts	43.5" x 17.75" x 0.75"	4" round cells	5 years	38 cells
	HE51	16.4 volts; 2.1 amps 34 watts	22.06" x 22.06" x 1.25"	85-mm square cells	5 years	square, high-density packing; 36 cells
	SX100	17 volts; 2 amps 32 watts	43.5" x 17.75" x 0.75"	4" square cells	5 years	square, semicrystalline cells; 40 cells
Solenergy Corporation	SG 1264-G	16.7 volts; 2.21 amps 34 watts	28.5" x 23.75" x 1.75"	4" round cells	5 years	36 cells
	SG 12664-G	15.5 volts; 4.2 amps 64 watts	48" x 22.9" x 1.75"	4" round cells	5 years	66 cells
SunWatt	H-150	17 volts; 8.6 amps 150 watts	95.8" x 47" x 8"	4" square cells	5 years	hybrid electric and water heater; 68 cells

*Note: This table is a comparison of like modules and does not represent all product offerings of all manufacturers.

In the past, manufacturers used a variety of methods to encapsulate the cells. Some embedded them in silicone rubber onto aluminum ribs, while others sealed them into glass or plastic cases. Now, however, in most modules the cells are sealed behind a tempered glass plate. The plastic used to encapsulate the cells, polyvinylbutyral, is the same material used in making automobile safety glass. It performs very well and will give a hermetic seal for decades. However, it has a temperature limitation below that of silicon, so the use of reflectors or fresnel lenses to concentrate more light onto the individual cells is restricted to manufacturers' recommendations.

Early modules experienced numerous failures due to stresses that pulled fingers off the cells or water that leaked in and corroded the connecting leads, but present-day solar cell modules are remarkably durable and long-lasting. An important design improvement is added redundancy. In a well-designed array, there are at least two connections between each cell, and sometimes more. The finger patterns allow the current to flow in multiple paths, and the connecting leads are soldered across the entire backs of the cells. These techniques increase the reliability of the module. If a cell cracks or a lead breaks, the performance of the overall system is unaffected. When planning to purchase modules, look for the features described—some small modules still have only one connection between each cell and could fail prematurely.

Making Your Own Modules

A good number of the people who use PV systems in their own houses have made the modules themselves. Some of these people acquired surplus cells and learned how to connect them. Others have attended Skyheat hands-on PV workshops where they assembled their own arrays, and many have successfully followed the instructions that have appeared in the media. *Alternative Sources of Energy* and *Cruising World* have published articles on making solar arrays, and the book *Practical Photovoltaics* details and illustrates step-by-step procedures for building several kinds of modules. Instead of duplicating those instructions here, we will simply refer you to these sources and add the reminder to make the array as rugged and dependable as possible. This can be accom-

FIGURE 5.8—Soldering copper connecting strips onto solar cells at a Skyheat
  hands-on workshop.   [David Ross Stevens photo]

plished by incorporating as much redundancy into the interconnections as possible, using long flexible wires or copper strips
across the entire fronts and backs of cells and sealing the array
securely.

It is now possible to purchase kits for the assembly of your
own modules. A note of caution—some kits are incomplete and
come with poor instructions for the inexperienced.

A practical application . . . .

PV power installation on an Indian reservation in the Southwest.
[Photo: ARCO Solar, Inc.]

CHAPTER 6
BATTERIES & STORAGE

Any discussion of batteries can easily become unnecessarily complicated. We'll try to keep it simple. Appendix G lists sources for further information.

A battery is an electrical storage device. Batteries come in many shapes and sizes, but all batteries have one thing in common —they store DC current for later use. In Appendix A we compare volts and current to pounds of water pressure and water flow per minute. We also compare resistance (ohms) to the resistance of hose size on water flow. To carry this water analogy one step further, we can compare a battery to a bucket. Just as we fill a water bucket for later use, so, too, do we fill a battery with power for later use.

There are batteries which can be used only once and then discarded, and there are reusable or rechargeable batteries (see Appendix C). We are concerned only with the rechargeable ones. But rechargeable batteries have important differences which make them suitable for different uses. For a PV system we need deep-cycle batteries.

A deep-cycle battery is one that can withstand repeated cycles of deep discharge and still function. A cycle is a complete sequence

FIGURE 6.1—A good example of a proper battery installation, with neatly done connectors, in a well-ventilated place.   [Photo: ARCO Solar Inc.]

of operation for the battery. It begins with a fully charged battery. In use the battery is "emptied" of its power. Then the battery is recharged or "filled up" again. Deep-cycle batteries are designed and made to handle this cycling.

An automobile battery is different. It is designed to give a car or truck starter a powerful, short spurt of electricity to get the engine turning and running. Then the alternator or generator takes over and recharges the battery right away. Auto batteries should not be left sitting partially discharged nor deep-cycled repeatedly. A new auto battery will soon die if deep-cycled.

Batteries are made up of cells, each rated at 2 volts DC. Thus, for a 12-volt system you will need six each 2-volt cells. Some batteries are an assemblage of cells, and you can find 6-volt batteries and even 12-volt deep-cycle batteries. Sears sells a 12-volt RV/marine deep-cycle battery and a 6-volt deep-cycle golf cart battery. A 12-volt PV system would require two 6-volt golf cart batteries.

A battery's size is an indication of its storage capacity just as a bigger bucket means more water capacity. A battery rated at 52.5 amps discharge per 8 hours can carry a load of 52.5 amps discharge for 8 hours before recharging is necessary. Golf cart batteries are typically rated at 200 ampere-hours capacity—you can run a 1-amp load for 200 hours. (Refer to Appendix A to change amperes to watts.) A typical auto battery is rated at about 50 ampere-hours and weighs 50 pounds.

Let's discuss the process of selecting the proper batteries for your PV system. First you must determine how many ampere-hours of storage your system will need to take you through cloudy periods and to handle overnight loads while also considering seasonal differences. Most areas have less available sunlight in winter, so it is a good idea to size your system for winter storage and loads. (Also refer to Chapter 4 for battery sizing.)

A battery bank of four each 6-volt, 200-ampere-hour deep-cycle golf cart batteries connected in series to give 12 volts and in parallel to give 400 ampere-hours of storage will provide about 8 days storage (400 ampere-hours battery capacity divided by 50 ampere-hours per day). Only two of the same 6-volt batteries would provide half the storage—about 4 days.

You can buy new batteries for your system or you can locate used batteries. When buying new batteries, be sure to tell the salesperson what your intended use will be and ask for suggestions. But, remember, few battery dealers are familiar with PV systems and all of them are out to sell batteries. Don't even consider a new deep-cycle battery unless it has a good warranty. Prices don't vary much, but it is a good idea to take the time to shop around for the best deal.

Buying used batteries requires special knowledge. You must check the battery appearance. Has it been kept clean and out of the weather? Does it show signs of physical abuse? Is the electrolyte clean and clear? Are the lead plates in good shape and not warped or brownish? Are there white deposits indicating sulfation? Are the terminals in good shape? Do the cells test properly

FIGURE 6.2—The interconnections between the batteries should be as short and as heavy as possible.

[Photo: ARCO Solar Inc.]

FIGURE 6.3—Electricity from the 660-watt PV array at North Arkansas
Community College is stored in this 2520-ampere-hour battery bank. The
C&D batteries are regulated by five ARCO Solar Village Power Panel
charge controllers, seen at the center right. The PV/battery installation
runs the lights and computer in the NACC Energy Center.

[Photo: Stephen Cook]

with a voltmeter or are they completely discharged? How were
they used and by whom? The telephone company usually takes
good care of its batteries and often disposes of them well before
their service life expires. On the other hand, warehouses and golf
courses may not have regular battery servicing programs, and
may wait until a battery is ruined before replacing it.

Now the details. A battery's self-discharge rate is typically 10 to
15% per year. This means that a battery will lose some of its
power just sitting around. Lead-acid with calcium or pure lead
grids have high (80%) electrical to chemical energy conversion and
typically have 1000 to 3000 deep cycles in them before they must
be replaced. But life cycle depends on the depth of discharge. If
you regularly dip deeply into your battery bank, you will not be
able to use them for as many years as you will if you just skim a
little power off the top and replace it daily. Automobile batteries
can handle a 10% depth of discharge and give you over 200 cycles.
Lead-acid PV batteries can handle 30% depth of discharge and give
you over 1000 cycles.

CALL NO. LAPTOP 2
    Date Due Slip

A-B Tech Community College
11/20/08 05:55PM

* * * * * * * * * * * *
* * * * * *

PATRON: 23312000444184

ITEM: 33312000016436
DUE DATE: 12/04/08
The solar electric home : a photovoltaics
how-to handbook /
abg
CALL NO. TH 7414 .D37 1983
            Date Due Slip

A-B Tech Community College
11/20/08 05:55PM

* * * * * * * * * * * *
* * * * * *

PATRON: 23312000444184

ITEM: 33312000016436
DUE DATE: 12/04/08
The solar electric home : a photovoltaics
how-to handbook /
abg

A note on safety. Low-voltage and low-wattage PV systems are relatively safe—it is hard to get a shock from a low-power system. But in that relative safety is a real danger. Carelessness can be the biggest problem. With a 12-volt system, you can put your fingers into a light plug and not get shocked—if you are dry. But the power in that plug can weld a ring on your finger. The current in your battery bank can melt metals—for example, a tool dropped across terminals. The acid in your batteries will burn your skin and eat your clothing—so be careful. When being charged, batteries give off hydrogen and oxygen. Under the right conditions, these gases will explode. Be sure to vent your battery storage area properly, whether it is in the house, under the house, or in a separate shed or room. When working on your system, disconnect your batteries by pulling a fuse, throwing a switch or opening a breaker. When working around the batteries, cover the terminals to prevent accidents from tools or metals shorting the cells or the system. Remember, the biggest danger with your PV system is that it is so safe that you might get careless. Some people put caution tags on their batteries. Such a reminder might be just the thing for you, too.

Servicing your battery bank requires only a few tools and a few minutes per year. You will need an hydrometer to test the specific gravity of the electrolyte. Your voltmeter will tell you the conditions of each cell or battery in your bank. Keep your batteries clean and out of the weather. Try to keep them about 70°F for optimum use. If they are located in an unheated room, keep them warm with insulation—not a heater! Make sure terminals and connectors are clean and bright. Coat them with mineral grease or petroleum jelly to prevent corrosion. Baking soda and water and a cloth will take care of spilled electrolyte and also remove dirt and dust.

A practical application . . . .

Owner-built PV/passive solar cabin in the Bull Mountains, Lavina, Montana.
[Design/Construction: Monique Mandali]

CHAPTER 7
REGULATORS

A regulator is a device used to maintain a desired constant output voltage and/or current regardless of normal changes in the input or to the output load. In a PV system a regulator is used to keep the batteries from being overcharged. Some regulators also prevent the batteries from being too deeply discharged by cutting off the load circuits and thus making it impossible to use any power until the batteries are recharged.

There are three types of regulators with which you should be familiar: (1) series regulators, (2) full shunt regulators, and (3) partial shunt regulators. Series regulators utilize a series transistor that electrically blocks the power from the array to the batteries when the batteries are fully charged. Full shunt regulators use a transistor to shunt or turn aside excess power from the array to a

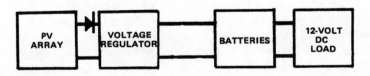

FIGURE 7.1—The wiring of a simple photovoltaic system using a regulator.

75

ground or dissipate it as heat or switch it to another load. Partial shunt regulators are used in larger PV systems and are actually a group of small shunts to minimize weight, cost and heat management (see Figures 7.2 and 7.3).

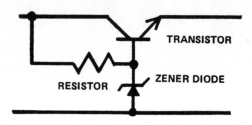

FIGURE 7.2—Series regulator.

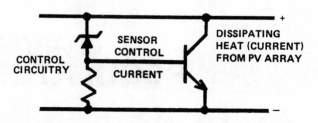

FIGURE 7.3—Shunt regulator. *Practical Photovoltaics* contains plans for building your own shunt regulator.

A good regulator will also have a trickle charge mode of operation. When your batteries are fully charged, the regulator will cut off or shunt most of the power being produced by the PV array. However, it will still continue to charge your batteries with a trickle or float charge, usually 1 to 5% of the battery ampere-hour capacity. This keeps your batteries "topped off."

Another type of regulator for PV systems you will be hearing more about is the switching regulator—an electronic device with a distinct advantage over series or shunt regulators in that it has low power dissipation and higher efficiencies (Figure 7.4).

Do you need voltage regulation? *Yes.* There's no doubt about it. You will need to regulate your power production to protect your batteries and ensure their long and useful life. The question is— what kind of control do you need? Just as there are different kinds of regulators, there are many ways to regulate. Let's take a look at them.

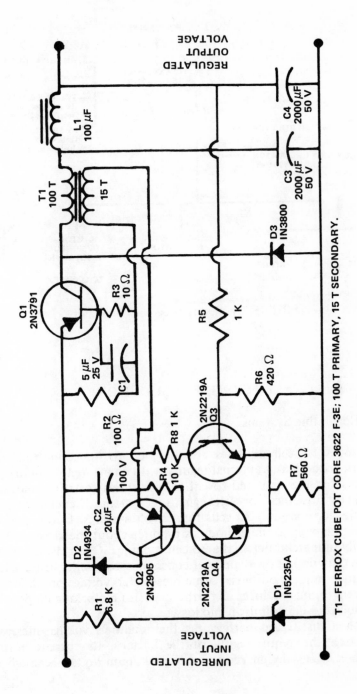

FIGURE 7.4—Switching regulator.

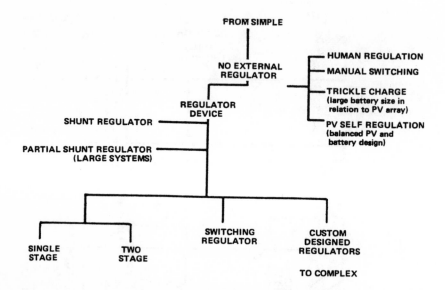

FIGURE 7.5—Regulator decision flowchart.

PV SYSTEM REGULATION

Self-Regulating Systems

Some solar cell modules are designed to be connected to a battery without any external regulator. In an effort to simplify PV systems for the user and to cut total systems costs, the industry offers PV modules for self-regulating systems. Generally, these modules have around 30 cells for 12-volt systems. During periods of heavy charging, these modules produce about the same amount of daily ampere-hours as the 33-cell modules. One reason for this is that they do not need a power-consuming controller or regulator to reduce module output at some preselected voltage point. Manufacturers' opinions differ as to the need for a blocking diode with its associated 0.5 volt drop and losses.

With a discharged battery, as the module's voltage increases to match the requirements of the battery, its current output decreases. This design reduces the maximum voltage enough to prevent overcharging the battery. As the module approaches its higher voltage range, current output is reduced to less than half, and the module becomes a trickle charger.

SunWatt, ARCO Solar and other manufacturers are now producing this type of module. The advantage is a reduction in module cost as it requires less material and less manufacturing per unit. In addition, balance of systems cost is reduced due to the elimination of the regulator. However, there are limitations to the use of these modules. They work well with small systems under 300 watts but require a large battery bank (75 to 100 ampere-hours per 35 watts).

Trickle Charge Regulation

If your PV array peak output current is less than 5% of the ampere-hour rating of your batteries, you do not need a regulator device. For example, if you are using one 35-watt PV module on your sailboat to keep your 100 ampere-hour deep-cycle marine battery charged, this small module can't possibly overcharge your battery.

Human Regulation

If you are operating on a limited budget, careful monitoring of your battery with a voltmeter and a good hydrometer can substitute for a regulator device. You can tilt your array to regulate the amount of power going to your batteries, or you can install a switch that completely disconnects your PV array. However, be sure to put up a large sign or reminder to tell you to turn your system on or off and to check your batteries frequently. Using a switch instead of a blocking diode is possible but a bad idea. Keep the diode in your system as it is a good form of low-cost insurance against nighttime and cloudy day discharging of your batteries through the PV array.

REGULATOR DEVICES

Most PV manufacturers, some distributors, and other companies offer proprietary regulators and most are interchangeable. Shop around for the one that will suit your special needs, both present and future. Windchargers also have regulators, but many of the older ones are not solid-state and use solenoids and relays which draw too much current from low-power PV systems. This is also true of automobile regulators. If you can get a schematic of a commercial regulator and are handy with a soldering iron, you can make your own regulator. Often, this is a good way to economize.

The ARCO Solar Village Power Panel (VPP) is dependable for regulation, and other firms make similar regulators. It is a two-stage series charge controller. That is, it provides full panel current to the batteries to the point at which they are nearly fully charged and then introduces circuitry that reduces the current to a trickle charge to finish the charge cycle. All wires come into and go from the VPP.

Power from the PV array (up to nine 35-watt modules) goes through the VPP circuitry to charge the batteries. When the batteries are nearly fully charged, a green light emitting diode (LED) comes on to indicate that the trickle charge is occurring. If the battery voltage gets too low or a system short occurs, the battery power circuit breaker trips and a red low battery LED appears.

FIGURE 7.6—Basic single-charge circuit control: Charge-a-Stat from Solar
Works!   [Photo: A. D. Paul Wilkins, Solar Works!]

These "idiot light" indicators, while helpful, do not give the whole picture, so you might add a meter box with ammeter and voltmeter. The VPP also has a terminal block for interconnections and three 5-amp load circuits with their own on–off switches. The entire panel measures only 6 x 6 x 3 inches and is wall-mounted.

The Solar Works! Charge-a-Stat does everything you need for regulation and more. In conjunction with the control board also available from Solar Works!, you can see what is happening with the system on the meters. The interconnects are easier to get to and each load circuit can be fused as needed using simple automotive fuses. The automatic switch makes it possible to charge an auxiliary battery bank or even make hot water with a 12-volt DC electric heating element. A PV system designed to give 100% PV power in winter often produces excess power during the rest of the year that would normally go to waste, but you can heat water, charge extra batteries, operate a fan or pump water once the main battery bank is fully charged.

Many manufacturers of regulators offer a wide range of options such as temperature compensation, stand-by generator startup, low-voltage warning lights or audible signal, meters (both dial and digital), circuit breakers and switches. Regardless of the claims made by manufacturers, regular testing of the specific gravity of your batteries with a hydrometer is essential.

A practical application . . . .

PV/grid-connect home in Chatsworth, California.
[Photo: ARCO Solar, Inc.]

CHAPTER 8
INVERTERS & AC

Simply stated, an inverter is a device that changes direct current to alternating current. For our purposes, we are talking specifically about 12-volt DC to 120-volt AC conversion (120-volt AC is sometimes called 110-volt AC, but it is all the same—the juice we get from the wall outlet in a grid-connect house).

The purist might try to operate a low-wattage PV system without an inverter—but there are problems and limitations to this practice. A few extra dollars for a 550-watt inverter opens up new horizons.

Inverters come in all sizes, shapes and price ranges, and offer many options. Some inverters produce a simple square wave form of AC which will power most motors. Other, more costly inverters produce a sine wave or modified sine wave form of AC and can power precision equipment and turntable motors, which require 60-cycle-per-second (hertz) AC power.

Inverters are either solid-state or rotary-type. The solid-state inverters are more efficient (70 to 95%) than rotary inverters (less than 85%). However, rotary inverters can handle power surges created by motor startups somewhat more effectively than the transistorized, solid-state inverters. The fact that motors draw about five times the rated wattage when first turned on must be taken into consideration to avoid burning out an inverter. In addition, inverters operate best when used at their rated capacity— you should operate a 300-watt load with a 300-watt inverter, not a 1000-watt inverter.

Some manufacturers offer built-in circuitry to handle motor startups, but that feature adds to system costs. Because inverters draw current when they are on but not in use, some manufacturers include automatic on–off switches, which also add to the price. Generally, those who use inverters for occasional loads put up with the slight inconvenience of manually switching on the inverter either at the unit or with a remote switch.

When installing an inverter, you must note polarity and keep the wires between the battery and the inverter as short or as heavy as possible. Size the wires to the inverter as you would car or truck battery wires—bigger is best. In any case, follow the manufacturer's recommendations on wire size and inverter operation.

Figure 8.1 shows a typical PV system with an inverter for the operation of 120-volt AC equipment.

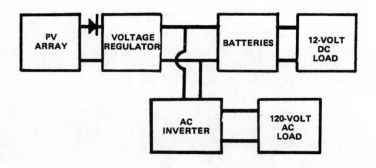

FIGURE 8.1—The wiring system of a typical solar electric home showing an inverter added to the system.

Appliances (120-volt) that primarily use resistance elements will operate well with almost any kind of inverter. In toasters, hair driers, slide projectors, and other devices that might have a small fan, the primary load is the resistance element or light bulb. Of course, the inverter must be capable of delivering the total wattage the device uses. 120-volt power tools with brushes will also work well on inverters; in fact, they will work equally well on 120-volt DC. Problems arise with solid-state inverters in powering induction or other brushless AC motors (such as refrigerator or washing machine motors). These motors interact with the inverter, producing distorted wave shapes and voltage spikes that can affect the operation of other devices plugged into the inverter at the same time, a particularly bothersome problem on appliance startup.

GRID-CONNECT

It is possible to install a large PV system and remain hooked up to the utility power grid. The advantage of such an installation is that the storage system for the home is reduced or eliminated. When the sun is not producing sufficient power to run a home's electrical appliances or the peak load is greater than production, the grid-connect PV home buys power from the utility company. When production exceeds consumption, the home is credited or actually paid for the power it produces for the utility.

A major manufacturer of inverters for grid-connect PV and wind power systems is Windworks, Inc. As of 1981 Windworks had supplied the Gemini Synchronous Inverter to over 900 homes. After a number of modifications over the past several years, the equipment has been endorsed by power company engineers. In fact, a utility company recently purchased the firm—lending to its credibility and encouraging greater acceptance of this technology.

The Gemini Synchronous Inverter is a line-commutated, line-feeding inverter which converts DC power to AC at standard line voltages and frequencies. In operation, all the available DC power is converted to AC. If more power is available from the DC source than is required by the home, the excess flows into the AC grid where it is used by others. If less power is produced than is being used, the difference is provided by the AC grid.

By interfacing with the conventional power lines, the need for storage is lessened. The inverter also has circuitry capable of handling an unregulated DC power input. For photovoltaic arrays, where the maximum power output is not a function of a single variable, automatic tracking circuitry seeks the highest output by incrementally varying the loading of the array while monitoring the power output.

The installation of such equipment is beyond the scope of the average homeowner, but the manufacturer has trained installers and works with local utilities to ensure a safe and optimum system design. When the amortization figures begin to look attractive,

FIGURE 8.2—This experimental house, with a very large array, is connected to the utility grid and will actually furnish power to the grid when output exceeds homeowner's needs. South view.   [Photo: MIT Lincoln Laboratory]

we will see thousands of PV/grid-connect homes. Until then, this type of PV system will be the exception and not the norm.

Another way of achieving grid-connect with PV is through the use of an induction motor as a generator. Ted Landers of the New Life Farm and now in charge of Perennial Energy, Inc. has long been using induction motors to cogenerate electricity with methane-powered internal combustion engines. The same procedure could be used with photovoltaics to power a grid-connect all-electric house. To quote from Ted's brochure:

> An induction generator electrically and mechanically is the same as an induction motor, except that it is driven (by an engine, water turbine, wind turbine, etc.) at a speed above the "no load" or synchronous rpm and at the same time is connected to the utility. The voltage and frequency produced will be exactly the same as the utility and the power produced will vary with the speed of the unit; the higher the speed, the higher the power output. Unlike a synchronous generator, it does not need expensive switch gear to feed power to the utility—just plug it in and run it. Also, if the utility goes down, it will stop generating and therefore is very safe and will not electrocute an unsuspecting lineman.

Such induction motors are available from several suppliers and through Grainger's catalog or distributors. Ted will also be able to help those interested in this form of grid-connect.

As more people make the PV/grid connection, there will be greater cooperation and familiarity on the part of the utilities. Although the number of people using alternate forms of energy to cogenerate electricity grows yearly, it is still unusual. Utility companies show some resistance in making it easy for the home-owner to make the grid connection, but increasing cooperation is expected as utilities become accustomed to the procedure. If you run into any resistance from your utility company, expect it in the form of uncertainty and a concern for safety. The PURPA regulations have definitely enabled us to gain greater control over interactions with utility companies.

A practical application . . . .

This PV array is affixed to the roof of an outbuilding to power an adjacent farmhouse in southeastern Michigan. After hoisting the array to its permanent position, the wires are connected to a junction box inside the outbuilding attic. [Photo: Alan D. Roebuck]

CHAPTER 9
INSTALLATION GUIDELINES

Now that you are familiar with the necessary components, we can look at putting the PV system in place. Each home is different and it is, of course, impossible to cover all possible arrangements and configurations. However, a few basic guidelines will help you think through your particular installation and enable you to avoid some pitfalls that cause initial installations to require modification and correction. As in all building projects, it is a good idea to go over the steps before actually starting the work. Write down each step and make notes to remind yourself as ideas and suggestions pop into mind or come from helpful friends and neighbors. While doing this, develop a materials list so that when you do start, all hardware, fasteners, wires and parts are on hand. If you live far from town, this is particularly important. An unfinished installation is an invitation to trouble. You can be sure that rain or snow will catch you at the most inopportune time.

The key rule to remember is: Keep It Simple. As your system changes with your needs, a simple straightforward installation that is easily accessible will facilitate modifications. And a straightforward installation will be a safe installation. Make all electrical connections in accordance with proper practices. Make sure all connections are mechanically tight and that you use the right fasteners for the job. Soldering requires some special skills developed with practice, but don't be afraid to try to solder and flux— you'll be surprised at how easy it is. In any case, wire nuts and fasteners will also work if done properly. Visit an electrical supply house and ask to see their wire connectors, stand-offs and terminal blocks. Become familiar with what is available in your area. Don't be afraid to ask questions. Most suppliers will be curious about your PV set-up and will want to help you get started.

The first step in the installation procedure is selecting the location of your equipment (review the section on site selection in Chapter 3). Be sure to read manufacturers' recommendations carefully; it is in everyone's best interest that the equipment they sold you be used to its best advantage. Most manufacturers provide free-of-charge installation guidelines and instructions. Some offer assistance by telephone should you run into any problems or have specific questions. Distributors and dealers will also help. It is also wise to buy your equipment from someone who actually uses the hardware themselves.

Site constraints require additional consideration. Since the electricity from the PV array is low-voltage DC, you must take into account the loss of power due to long wire runs from the array to the home. In general, a pole mount and heavy wire are better investments than cutting down a shade tree. However, there are limits to this sort of thinking. Unless you can get a good deal on new or used cable, or your wiring budget is unlimited, there is a point at which you have to compromise in site selection. Check the wire size chart (Chapter 10) and local wire prices to determine the most practical distance between your array and home. And, while you are at it, think about unsightly and unsafe overhead wire runs as compared to underground cable.

Accessibility to the installed equipment is important in siting the array, the battery bank, and the regulator/control board or box. A short wire run to an unhandy niche in a dark, crowded corner is false economy. If you can't move around freely in the battery and controls location, anything from accident and injury to plain aggravation will make you wish you had spent the extra dollars for longer and heavier wire to that open access spot. I mounted my first regulator on a wall in the corner behind a table. Each time I changed the system to test new methods and products, I had to move furniture, grab a flashlight and a mirror, and work upside down in a tight place. Eventually, I wised up and made the controls and junction box more accessible.

As always, safety first! Be sure your array and batteries are disconnected while working on your system. See Chapter 6 for a complete discussion of battery location.

Locating the regulator and controls is equally important. Some regulators require proper ventilation to keep the internal parts cool. Dirt and dust on controls and contacts can be a problem, too. If your regulator must "breathe," isolate it in a dust-free area. But don't put it in or near a closet where someone might use it for a clothes hanger. Hardware stores sell electrical boxes for equipment mounting and protection. Never locate your control box and regulator directly above a heat source.

A note on mounting. Simple temporary ground mounts are suitable for testing purposes—just legs attached to the array frame. However, you will be taking a chance with such a set-up if you

FIGURE 9.1—Typical pole mount structure. Some manufacturers use a horizontal framing member at the base of the array, but that eliminates the seasonal adjustment.
[Photo: Solavolt Internat'l]

[Solavolt International is a partnership of Solar Energy Inc. (a subsidiary of Motorola Inc.) and SES Incorporated (a subsidiary of Shell Oil Co.)]

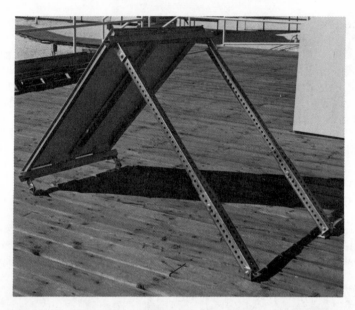

FIGURE 9.2—Typical ground mount structure. Note the adjustable rear legs for seasonal alignment.   [Photo: Solavolt International]

FIGURE 9.3—This owner-built ground mount structure shows simplicity in design and ease of construction. Especially with small arrays, it is easier and less expensive to make your own mount.

[Photo and design: Bruce Wheeler]

leave it too long. We have experience with both fixed and move-able roof mounts. In our installations, the fixed mounts were attached to wooden or metal stand-offs of 2 x 4 ft pressure-treated lumber blocks or channel steel or aluminum bolted to the roof rafters. Then the array was fastened with bolts to the stand-offs. The moveable array was mounted in the same way except that metal legs from the sides of the array were added. These legs made seasonal adjustments possible. Be sure to look at a few billboards and signs on stores to see what kind of metal frames, angle iron, tubing and hardware is used in your area. If a PV array were to tear loose in a wind storm, it could be destroyed and might hurt someone in the process. Roof-mounted signs and arrays have safely withstood the elements for decades. Try to maintain this good record.

Seemingly obvious, but of utmost importance for roof mounts, be sure to fasten your array securely. The best way to check the array mount is to grab it and pull. Hang on it if you must to test the strength. A periodic inspection of the fastenings is a good idea.

Don't forget to protect your batteries, open wires and fasteners from the possibility of falling objects. Children and the curious will be all hands. Keep things out of reach, safe and secure.

A practical application . . . .

A 10-watt panel trickle charges the battery at this remote weekend cottage in Norway.   [Photo: Solenergy Corporation]

CHAPTER 10
WIRING PV SYSTEMS

SAFETY FIRST!

Wiring a PV-powered home is the same as any other wiring job when it comes to safety. It is very important that you follow proper wiring practices to avoid shocks, fires and other hazards. If you live in an area where building inspections are required, consult with the local authorities. Living in a wilderness area, however, does not free you from safety concerns. In fact, remote home dwellers' freedom from the interference of building inspectors must be tempered with the responsibility of providing for their own safety. Sloppy, inadequate or unsafe wiring is an invitation to tragedy.

Safety is based on knowledge and knowledge rests on a foundation of fundamentals. There are many easy-to-understand instruction books on conventional home wiring. Once you know what is required for the safe wiring of a conventional utility-powered home, you can modify available equipment to serve your special needs. Sears and Montgomery Ward have good, inexpensive how-to booklets for the do-it-yourselfer, and all public libraries have home wiring books. *How to Be Your Own Home Electrician* by George Daniels is a favorite reference.

WIRING HOW-TO

The following information has been provided by Steve Willey. Steve lives in Idaho, in a unique home powered by two wind-chargers and photovoltaic panels. He has extensive electronics knowledge and practical experience, and teaches classes in energy conservation and solar energy. Professionally known as Backwoods Cabin Electric Systems, Steve provides information, design consultation and equipment for the home alternative energy user.

How to Get It Together

The easiest way to connect your photovoltaic panels to your batteries is to use the meter and fuse box like the one produced and sold by Backwoods Cabin Electric Systems (Figure 10.1). You just hang it on the wall and connect to it the wires from your solar panels, batteries and five house wiring circuits. The interconnections described here are already done within the box, and the parts mentioned are mounted on the front. The box has a solid varnished 2 x 4 frame 18 inches wide, 10 inches high and 4 inches deep. One meter shows how much current is coming in from the solar panels; another shows how much current is being used in the house. There are switches to shut off either current. The diode necessary for photovoltaic panels is already hooked up on the meter box and can handle up to four panels. A diode that

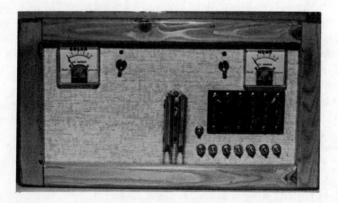

FIGURE 10.1—Steve Willey's meter fuse box for PV or small hydro (up to 15 amp) use.   [Courtesy: Steve Willey; Orion Renk photo]

will accommodate more panels is available. There is an indicator to test the diode and another to let you know if anything electrical has been left on in the house when you go out. Another option is a connection to recharge flashlight or radio rechargeable ni-cad or gel-cell batteries from your house batteries (see Appendix C). On the front panel are six snap-in fuses—one for the battery and five for power to your household lights and outlets.

The circuits involved are really pretty simple. The next sections will explain how to set up your own panel-to-battery connection, and how the parts work.

Equipment Availability and Specifications

Wire. For normal runs (35–40 feet) to your rooftop PV array, you can use No. 10 Romex wire when connecting up to four panels. Secure the wire to the mount to prevent its weight from pulling against the terminals on the panels. There should be some slack in these connections to avoid strain. Use single, separated pieces of this wire to connect individual panels, making note of the color used for positive (+) and negative (–) so that you will connect it correctly at the other end. If you have more than four panels, you can use a second Romex No. 10 to bring down power from the second group of one to four panels. Or you can use No. 6 wire, but in that case you must splice it to smaller wire to connect to the panel terminals, so it will be flexible enough. Be very careful not to reverse the battery connection during installation. That would result in the battery and panel pushing in the same direction in the charging loop, and too much current would flow for the survival of the panel. Observe all + and – markings on panels, batteries and diagrams!

Diode. Each diode has a rating of the maximum current it can carry in the forward direction and maximum voltage it can withstand in the opposite direction where it will not pass current. Most will withstand 50 volts or more. I use 200-volt rated diodes for a safety margin since the cost is only a little more. Diodes come in current ratings of less than 1 amp to 500 or more. Select one rated for more current than you ever plan to run through it. I use 12-amp diodes for up to 8- or 10-amp charge circuits, and 55-amp diodes for above that. The 12-amp diodes must be on a heat sink surface about 12 inches square. Try the power transistor heat sink and heat sink compound available from Radio Shack.

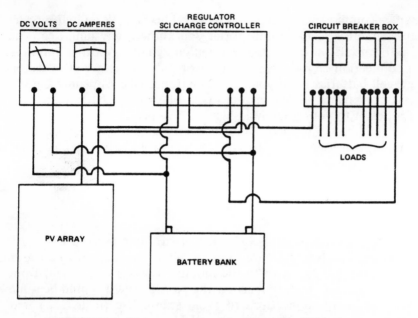

FIGURE 10.2–Wiring graph diagram and circuit diagram.
[Courtesy: Specialty Concepts]

Table 10.1–Wire Resistance
(ohms per 100 ft @ 68°F)

Gauge	Copper	Aluminum
000	.0062	.0101
00	.0078	.0128
0	.0098	.0161
2	.0156	.0256
4	.0249	.0408
6	.0395	.0648
8	.0628	.103
10	.0999	.164
12	.1588	.261
14	.2525	.414

Meters. Meters cost from $10 up to $50. I use the less expensive ones which are accurate enough for battery charger needs. Automotive ammeters with 0 center are impossible to read at 2-, 4- or 6-amp charge rates. Their scale ranges from 0 to 30 or, more often, 60 amps in half the scale width.

Fuses and Holders. Auto parts stores have fuseblocks which hold one to six fuses each and have screwdriver connections to the No. 12 Romex wire. Use AGC-type 10-amp fuses for each house circuit.

Wiring a 12-Volt Home

After establishing a working power source (*e.g.*, wind, solar, hydro, gasoline) and obtaining the proper type and quantity of batteries, there are three more things you must do.

1. *Wiring.* To complete the circuit, wire inside the house to bring power to the point of use, and wire from the battery to the source of power.
2. *Connections and Fuses.* In an orderly manner, connect all wires to the battery terminals, to provide not only a safe mechanical connection but a means of metering the power from the source to the battery to verify that it is working. Metering of power being used is necessary to determine if you are using more or less than your system can supply continuously. Fuses for each wire are needed to prevent a house fire in the event of wiring failure.
3. *Switches, Outlets and Lights.* At the ends of the wires— from the battery to the point of use—you will need to install switches, outlets and lights, as well as your chosen appliances. The lights and outlets, particularly the outlets, must be selected with DC in mind.

Wiring Types Available

To carry the current flow at 12 volts, wire has to be larger than the wire used in 120-volt houses. It is assumed that since you conserve energy, you can get by on the wiring used in a 110-volt house because your power requirements are less.

If the wire used is too small, or too long, the current flow in it will be restricted. Using just a small amount of current, the flow may be fine, but when the current increases, such wiring will restrict the larger flow—lights dim when other lights are turned on.

To avoid this, use the proper size wire. No. 12 Romex is standard house wiring and is the best bargain. No. 10 is slightly larger,

No. 14 smaller. Any will work, but the larger the better for long runs or for higher current uses (motors or inverters). Romex is usually three-conductor, having a black, white and extra safety ground wire. If you are not wiring to code, you may find Romex without the extra ground at bargain prices since contractors no longer buy it. Lamp cord, which is easier to work with, is considerably smaller (No. 18) and costs as much per foot as Romex. It is *not* recommended for permanent wiring.

Meters, Fuses and Interconnection

It is best to use a separate run of Romex to each area of the house. If wiring to the electrical code standards, follow the rules. Otherwise, you can supply several outlets and lights from one Romex run from the battery, *if* the outlets are fairly close together, and *if* the total distance is not over about 25 feet, and *if* no use is planned for over 5 amps. If you use the smaller No. 14 wire, make that 15 feet.

Start at the battery area and run wires as directly as possible to each location rather than looping a long route around the house. At minimum, run wire half way around the room in one direction, and another wire half way around in the other direction. In multistory buildings, it is easy to run wire straight up a wall to an outlet and light switch on each floor.

Where all wires come together, you need to connect them to the battery and provide a separate fuse for each and a current meter (ammeter) for the total.

Figure 10.3 shows how to connect a set of branch circuits to a 12-volt DC storage battery. The following parts are needed.

1. Automotive-type fuseblock ECHLIN FB6260 from NAPA auto parts.
2. 5-amp fuses (circuit breakers for 120-volts won't be accurate at 12 volts and cost more to mount—not recommended).
3. Ammeter 0–10 or 0–15. Auto-type reads –30– 0–+30 and will not be sensitive enough—not recommended.
4. Standardized outlets. Polarity is important.
5. Switch. You may want a main switch to turn all power off.
6. No. 10 automotive-type wire. For connecting the common + and – to the battery. Be especially careful when wiring from the battery to the fuseblock, as there is *no* safety fuse.

7. The negative wires must be fastened together to a wire to the negative battery post. Use a copper strip with a row of screw connectors (available from any supply store) or a ¼-inch bolt with washers between each wire. First mount it to the plywood with one nut, then the wires and washers, and then the second nut. Do not bunch all the individual wires on the battery terminal.

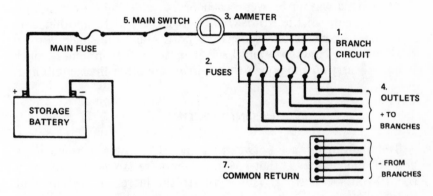

FIGURE 10.3—Simplified wiring diagram for batteries to fuse/control box.

Make all connections neatly, stripping no more insulation than is required to make the connection. Be sure that no wires can move and touch. All these parts can be mounted on a 1-square-foot piece of plywood. A larger piece of plywood can accommodate the meters for your charging system, too.

Connect the positive wire of each Romex run to the bottom screw of each fuse; that is your positive power source to that Romex run. If there is an accidental short circuit (+ and – wires touching), power will try to flow through the wire from the battery in almost unlimited quantity (amperes). Only the slight resistance in the wire limits it. The wire gets hot, sometimes red hot, and can cause all sorts of problems. However, since you just wired-in the fuse, any flow of current in excess of 5 amps will melt the fuse immediately, and the power is thereby disconnected. Your peace is preserved! It takes over 6 x 5 amps at 12 volts current to get Romex No. 12 hot. Always use fuses in the line, and keep spares around so you aren't tempted to replace a blown fuse with something that doesn't limit the current potential—like nails and bolts.

The negative wires from the Romex must all be connected to each other and to the negative post of the battery. An electrical supply store may be able to sell you a copper strip with a row of screws or clamps on it for connecting a number of ground wires. If so, great. If they say they don't know what you are talking about or they will get it in a few weeks, there is another way. You can get a ¼-inch bolt, several inches long, *threaded all the way*. Then, with a washer between each or by soldering on wiring tips, you can clamp all the Romex wires together and to another wire that goes to the battery's negative post. Make the single wire to the battery a larger size and keep it under 3 feet in length, since it has to carry *all* the power that is in *all* the other lines combined.

CONCLUSIONS

The main difference between wiring a PV-powered home and a grid-powered home is that PV homes require larger wire to carry the 12-volt DC current. This is due to the increased resistance at lower voltages.

When wiring a PV-powered home, you must take into consideration the voltages and currents being used in each circuit. It is wise to plan for the maximum loads so that your wiring will be able to carry the power you plan to use in the future. The initial cost of the wiring will be offset by the flexibility of the system as your needs change and as you expand your production and consumption of solar electricity.

The wiring from the PV array must be sized to carry the peak current from the array to the battery bank or regulator/control board. The wires from the battery bank to the individual appliances must be sized to carry the appliance load and must be properly fused for protection against short circuits in the wiring or the appliances.

Special outlets, switches and other electrical wiring equipment are available from recreational vehicle suppliers and some of the suppliers listed in Appendix G. This special equipment works well but is not absolutely necessary. You can use standard, inexpensive outlets, plugs and other readily available equipment if you adapt them to your special needs.

Polarity is critical with a direct current system. If you decide to use standard outlets, be sure to wire *all* the outlets in the house in

FIGURE 10.4–Simplified schematic of house wiring showing both 12-volt DC and 120-volt AC outlets, as well as several sources of electric power.

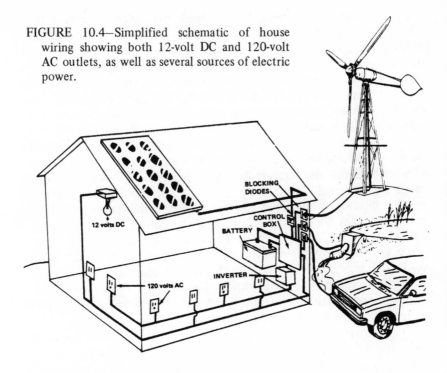

the same way. That is, fasten the white wire onto the same terminal at each outlet, likewise with the black, and then place a red dot to indicate the positive side of the outlet. Do this with plugs, too. Thus, you will have both polarity and uniformity.

Many people use cigarette lighter-type outlets and plugs to ensure proper polarity and avoid accidents. Some plugs and outlets have large and small slots, and others have configurations that make it impossible to reverse polarity. These can be used, but they cost more than the conventional type.

An important consideration in the selection of outlets is possible confusion to visitors in your home. If there is a possibility of an accidental reverse polarity when plugging in an appliance, then it is best to prevent such an occurrence by using the commercially available DC outlets and plugs. In addition, if you have an inverter and 120-volt AC outlets, be sure to make it impossible for someone to plug a DC appliance into your AC outlet and vice versa. Some outlets have covers over them (outdoor-type or child-proof) which will help avoid mis-pluggings.

Switches are no problem. Some people use automotive switches, which are specially designed to carry DC current; others use standard house switches. However, when a switch in a DC circuit is opened and closed, the contacts will arc, causing eventual deterioration of the contacts. To avoid this, you can put a capacitor across the switch–wire connections. PV equipment suppliers and electronics supply houses can help you select the appropriate capacitor. To avoid or lessen the switch contact arcing problem, use the older-style switches—the loud clicking (less-expensive) ones. Do not use the "silent" mercury switches.

Some people use circuit breakers for their PV wiring. Those made by Mechanical Products Inc. seem to hold up well and trip reliably in tests conducted at Skyheat. Circuit breakers are automatic switches, and even the best have the contact arcing problem. For this reason, most 12-volt DC homes use automotive fuses. They are inexpensive, readily available, offer more flexibility in system sizing, and can often be found for almost nothing in salvage yards. If you wire a room for the eventuality of larger loads, you can start with small-value fuses and change them as the load increases to ensure a safe circuit at all times.

CHAPTER 11
TESTING & MAINTAINING
YOUR PV SYSTEM

In the normal course of events you will hardly notice your PV power system. You will use your lights and appliances and tools and feel a mild amazement at its quiet, dependable power. After a while you will tend to take your PV power system for granted. If a problem occurs, however, it should be no surprise because you have been keeping a relaxed eye on things.

The Solar Power Corporation (Exxon) Installation and Maintenance Manual states: "Properly installed solar electric generator systems should only require regular maintenance visits once a year." You'll probably want to check out your installation and keep an eye on things more often than that to become comfortable with your new set-up, but that should only be to satisfy your desire to get to know this new gadget in your life. PV systems actually work better if left alone.

Test your modules when you first receive them. Prior to testing, inspect all equipment for manufacturing defects and shipping damage. File all papers, receipts and warranties in a safe place.

You need only a few tools for testing. A good multitester and a good hydrometer are essential. You will also need a load resistor. The multitester will be used many times to test AC and DC voltage,

AC and DC current, resistance and continuity. I use a 20,000-ohms/volt multitester with a 10-amp DC scale. If your system is large, you will want a separate DC ammeter (0.1 volt drop across meter). Don't be cheap—buy good meters that will last a lifetime, and take care of them.

The load resistor will be used to test each PV module for output under simulated in-service conditions. I use a 10-ohm wire-wound ceramic resistor capable of handling the 35-watt capacity of my ARCO Solar PV modules. Use the Ohm's law chart in Appendix A to determine the size resistor you will need for testing. Resistors can be purchased at electronics supply houses or you can make a resistor with a length of wire of the proper resistance (as determined by the multitester). Some older cars have adjustable resistors in their starting circuits or variable-power resistors in their heater fan controls.

Now for testing. First, test your PV modules' open-circuit output. Measure the voltage and current at noon on a sunny day with no haze. Record your readings for each module. Next, put your load resistor across each separate PV module's positive and negative lead or terminal. Voltage is measured parallel to the load; current is measured in series with the load. Record your results. These initial tests are important if your PV modules have a wattage degradation warranty. Readings should be at least 70% of manufacturer's rating for current and 80% for voltage. Date each test result sheet and keep them with your records.

Test your batteries with the hydrometer following manufacturer's instructions. (Be careful not to drop any battery acid on yourself or your clothes.) Measure the specific gravity of each cell once a month during the first year. Your batteries should be fully charged initially. You may want to test your batteries for self-discharge. This test takes a week and gives a good indication of the useful life remaining in used batteries.

To self-discharge test, disconnect the batteries and charge them with a standard automotive charger at a current rate not exceeding the battery's capacity in ampere-hours divided by 20 hours. For example, a 200-ampere-hour battery should be charged at 10 amps or less current. When all cell voltages are 2.3 volts, the battery is fully charged. Record specific gravity readings for each cell. Keep the batteries out of use for a week at room temperature and test for specific gravity again. Good cells will have a self-discharge rate of less than 0.015.

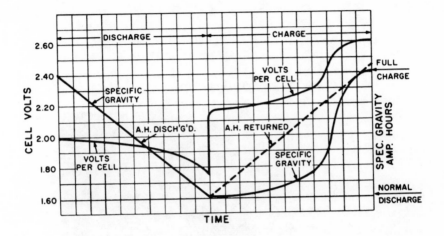

FIGURE 11.1—Typical voltage and gravity characteristics during a constant
rate discharge and recharge.   [Courtesy: Exide]

Next, test the entire PV system. Measure the current and voltage
output. Test and record voltage and current—both open-circuit
and with your regulator and/or batteries in the circuit as a load.
You should probably test the entire system several times the first
year; after that once a year is sufficient.

Maintaining your PV system is very simple. Check your battery
electrolyte level and replenish with distilled water as necessary.
Don't overfill. Clean and tighten battery posts. Keep your batteries clean since dust and grime can conduct current between the
positive and negative terminals and the case. To retard corrosion,
put a little petroleum jelly or other protectant on the posts and
connectors.

Check your PV array for dirt buildup. If you live in town or in
a pollution zone or on a well-traveled road, you will probably have
to wash your array one or more times per year. Arrays with 15°
tilt usually stay clean with rain. When you do clean your array, use
a mild soap or plain water and a soft cloth. Do not use solvents or
strong detergents.

Once a year check all appliances, wires, cords, connectors, plugs
and outlets. Repair or replace as necessary.

A practical application . . . .

Ground-mounted PV array in California.
[Photo: ARCO Solar, Inc.]

CHAPTER 12
HYBRID ELECTRICITY & HEAT SYSTEMS

Solar cells operate more efficiently if they are cool. On a summer day, a roof-mounted photovoltaic module can easily reach temperatures of 150°F (65°C) and higher. The cells will work a lot better if some of that heat is removed. If any sort of concentration system is used to increase the amount of light falling on a set of cells, provision for cooling becomes a necessity. Some manufacturers have built cooling fins into the backs of their modules. It is also a good idea to mount the array on a rack to increase air circulation.

However, it is a shame to throw all that heat away. This is where the hybrid module comes in. A hybrid module removes the waste heat from the solar cells and uses it to produce either hot air or hot water (or possibly both), as well as electricity, and has a number of advantages over standard solar cell modules.

ADVANTAGES

Greater Electrical Efficiency

The main reason for utilizing a hybrid module is to cool the solar cells. The drop in efficiency of a hot cell shows up as a lower

109

output voltage; the short-circuit current of a cell actually increases slightly at higher temperatures, but the open-circuit voltage drops off nonlinearly. This means that the loss in cell (and array) voltage is minimal as the temperature goes up to about 110°F but increases dramatically as the temperature continues to rise.

If a module is charging a relatively dead battery, no loss in charging current will result until the module gets above 150°F (65°C) or so. A cool array could operate more efficiently but wouldn't under these circumstances since the voltage of the system is determined by the battery voltage, and the charging current is determined by the sunlight intensity. As the battery approaches full charge, the system voltage rises and the hot module will operate less efficiently, causing the charging current to drop off faster than it would for a cooler module. As long as the module is kept below 130°F (55°C), the actual performance in a home system will be satisfactory. There are times when a hybrid module will be running at a much higher temperature than a conventional PV module, especially in the winter when the incoming water is already 110 to 120°F and the system is still producing heat. If the batteries are stored outdoors and cold, the lower output voltage of the hybrid module may cut the charging rate considerably. Storage batteries need a higher charging voltage when they get cold and should be kept in a warm, sheltered environment if at all possible (also see Chapter 6).

For hot water, 130°F is a very good temperature. Since most remote homeowners intend to have a solar hot water heater, either as their main domestic hot water supply or as a supplement to a wood or fossil fuel heater, considerable expense and trouble can be eliminated if the water heater and electric generation systems are one unit. The compact hybrid systems are also an advantage if roof area with good solar exposure is at a premium.

Greater Thermal Efficiency

Not only is a hybrid module a better electric generator than the equivalent single-purpose module, it can also produce more hot water than the conventional solar water heater. This was the surprising result of an experiment at Skyheat where a prototype hybrid system was mounted side-by-side with a same-sized standard solar water heater. The heater, built from a Suncor kit, has an absorber made of aluminum fins painted with a nonselective black coating. The water passes through a single copper tube fastened to the back of the fins in a serpentine pattern. The single transparent cover is made of low-iron tempered glass.

The SunWatt hybrid prototype shared many of the kit's components: the cover glass, frame and case are all made with kit parts. The hybrid case is twice as deep to accommodate the reflectors installed to increase the light intensity on the cells. Figure 12.1 shows the two systems mounted on the roof of the Skyheat workshop. We plumbed the collectors with control valves and flowmeters so that the incoming water was split into two streams (see Figure 12.2). This allowed us to have identical flowrates in the two slightly different pipe systems so an exact comparison of the temperature rise and efficiency was possible. The

FIGURE 12.1—The hybrid PV/hot water solar collector shown in the background produced more hot water than the identically sized conventional solar water heater mounted next to it.   [Skyheat photo]

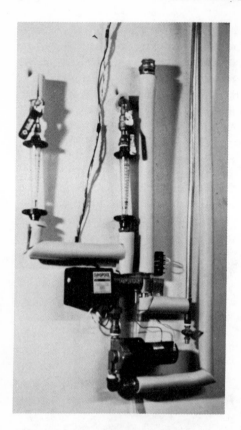

FIGURE 12.2—The control and
plumbing system for the hybrid
PV/hot water experiment. The
draindown valve and the pump
both operate off the 12 volts
DC produced by the hybrid
module. No external control is
used.   [Skyheat photo]

hybrid module operated more efficiently than the standard heater.
This high thermal efficiency seems to be independent of the
amount of electricity being drawn from the cells. One possible
explanation for this appreciated but puzzling synergistic behavior
is the fact that silicon solar cells are a rather good selective surface.
In fact, it has been proposed that silicon or some other semiconductor be used as the absorber surface for high-temperature solar-
thermal systems.

DISADVANTAGES

There are some disadvantages to the hybrid modules and some
potential trouble spots.

When the hybrid system at Skyheat starts up in the morning, it gets very hot before the photovoltaic cells generate enough current to start the circulation pump. When the pump does start working, it sends a surge of cold water into the hot fins. So far, this thermal shock, repeated every day, has had no effect on the performance of the device, but the interconnect wires must be carefully laid to absorb the thermal expansion and contraction inherent in the system.

Since a hybrid system has plumbing or ductwork connected to it, it is much less portable than the usual PV module. This means that any scheme to move the array and track the sun is much more complex and prone to leakage or breakdown. One of the hallmarks of a photovoltaic system is its reliability and freedom from moving parts, so a hybrid installation should be designed so that it need not be moved, keeping it simple to use and maintain. The Winston concentrator system used at Skyheat will accept and concentrate both the winter and summer sun with no adjustment. This is geometrically possible with a concentration ratio of 2 to 1 if the case is tilted at an angle equal to the latitude.

Higher concentration ratios are possible, but the cells *must* be cooled. This means that the heat extraction method must be very reliable, for if it fails, the cells—encased in an insulated box and absorbing all those watts of sunlight—can get hot enough to melt the solder on the interconnecting wires. The purpose of the hybrid module is to use all that heat.

AVAILABLE HYBRID MODULES

The only commercial hybrid module available today is the Sun-Watt H-150, but Solar Technologies, Inc. may again soon be manufacturing a PV/hot air module called the SunLoc III. This module has proven successful in several installations in the east.

Both modules use aluminum reflectors to concentrate more sunlight onto the cells. The higher concentration ratio in the Sun-Loc III requires that some sort of tracking device be used. This mechanism is built into the case so that installation is relatively simple, but the added complexity raises the cost and the possibility of a malfunction. However, each cell is more fully utilized and works harder than in the SunWatt module, where the concentration ratio of only 2.1 to 1 allows a simple fixed-cell position.

INSTALLING HYBRID MODULES

Hybrid systems are a little more complex to install than the usual PV array. If concentrators and/or any kind of tracking system is used, the system must be accurately positioned facing due south and at the angle designed into the unit for your latitude. An error of only a few degrees may degrade the performance markedly.

The hybrid module cases must be securely fastened to the roof or to substantial support brackets. The units are big and a strong wind can put a real strain on the mounts. For extra support, go right through the roof rafters with lag bolts or build a backup plate in the attic.

Plumbing or ductwork connections must be planned also. It is helpful to have some experience installing solar water or air heaters. The plumbing connections to the SunWatt module are made in the same fashion as to any manifold-type solar water heater. The individual fins are mounted at a slight slope in the case so that they will drain completely when the top of the case is level. This particular hybrid design must be installed long-side horizontal. Copper pipe can be soldered directly to the inlet and outlet, or short lengths of rubber hose can be clamped onto the connections. The hose should be of high quality, able to withstand both the water pressure and the environment.

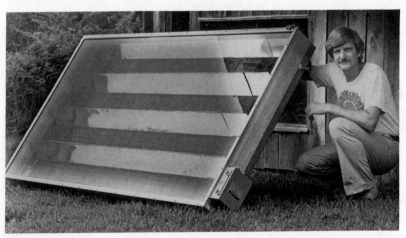

FIGURE 12.3—The SunWatt H-150 hybrid photovoltaic/hot water module incorporates trough-type reflecting concentrators to produce 150 watts (electrical) and 1600 watts (thermal) at the same time.   [Bill Bishop photo]

BUILDING YOUR OWN HYBRID

It is possible to convert some conventional solar cell modules to hybrid systems, but this is a very tricky business. Many of these modules, such as those manufactured by ARCO Solar and Solarex, are made with thermoplastics that cannot withstand very high temperatures. You can't add reflectors to these units to increase their power output except in a limited way. If you were to build them into cases to use them as solar air heaters, they could be wrecked by the stagnation temperatures reached should the forced air blowers quit. Some older PV modules were made by encapsulating the cells in silicone onto finned aluminum extrusions. The backs can be closed to make an air duct and, if no reflectors are added, the stagnation temperatures should do no harm. (The commercially available hybrid systems are made with materials capable of taking the greatest expected stagnation temperatures.)

Another way to build your own hybrid is to purchase just the cells and construct the entire module. *Practical Photovoltaics* contains step-by-step illustrated instructions. The aluminum "Big Fins" sold by Zomeworks or those used in Suncor water heaters can serve as a base to hold solar cells for hot water systems. While it is not difficult to make your own hybrid fins, we recommend strongly that you attend a Skyheat workshop, if possible, or follow the instructions in *Practical Photovoltaics*. These procedures were developed after several years of experience and ensure a long-lasting system.

INTEGRATING HYBRID SYSTEMS
INTO PASSIVE SOLAR ARCHITECTURE

For a completely self-sufficient home, the hybrid PV/hot water system is an ideal adjunct to a passive solar dwelling. If provisions are made in the design to incorporate the system in the building structure, it is possible to furnish all needed electricity and hot water with a minimum of piping or control systems.

If the hybrid fins are installed inside an attached greenhouse or sunspace, a complex freeze protection system is unnecessary. A simple arrangement to close off and completely drain the collector and related pipes will take care of those few times when you might be away in the dead of winter and the sunspace may get below freezing. Using plain water in the system makes possible a very simple plumbing system.

Figure 12.4 shows a water supply system used successfully in several southern Indiana homes. The water is rain water from a cistern, but a well would also work. There are a variety of 12-volt DC powered pumps capable of pumping water 30 feet or more. The pump does not need a large capacity as it can take longer to refill the tank than it takes to draw it down. A simple float switch turns on the pump when the tank gets low. If you are fortunate enough to have a spring uphill from the house, a float valve or an overflow pipe can replace the pump.

Positioning the hot water storage tank lower than the cold water tank ensures that the hot water tank is always filled. Be sure that the highest point in the hot water plumbing has an air vent valve and is below the low water level in the cold tank at which the float switch turns on the pump. If the hybrid module is below the

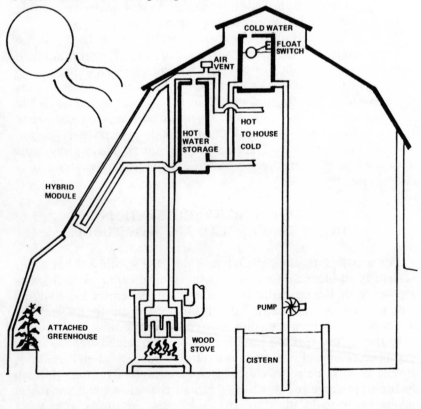

FIGURE 12.4—This gravity-feed water system uses very little electrical energy to supply hot and cold running water. The single pump can be PV-powered.

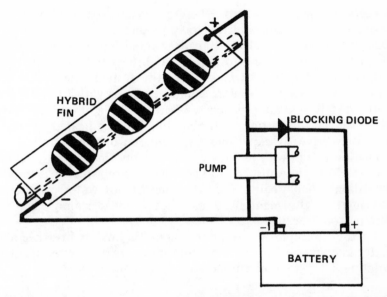

FIGURE 12.5–How to connect a hybrid module to both its circulating pumps and a storage battery. When the pump is running at full speed, the excess current is stored in the battery.

storage tank, efficient thermosyphon action will circulate the water as the sun heats it. Zomeworks and Sunspool offer low-resistance check valves that will prevent backward syphoning at night, allowing you to place the collectors and tank at the same level, if necessary.

If the collector must be higher than the storage tank, a circulating pump is needed. Of course, this pump can be directly powered by one of the photovoltaic cell strings in the hybrid module, a control being unnecessary. The pump will run only when the sun shines. Figure 12.5 shows the wiring on such a system at Skyheat. In this case, the pump only draws 1 amp at 12 volts when running at full capacity, while a single fin can produce almost 2 amps in direct sunlight. The excess power is fed to the storage batteries. The blocking diode keeps the battery from powering the pump when the sun goes down.

The gravity-fed systems installed at Skyheat use woodstove heat exchangers as backup hot water sources in the winter. Again, these can be thermosyphon systems.

If a solar cell array is placed in a greenhouse, remember that only one cell need be shaded to reduce the total output of the series string nearly to zero. Therefore, you must be very careful about the placement, and that may require some creative design. One fellow in Kentucky installed a whole line of collector fins (with reflectors) on the roof ridge of his greenhouse. The array works very well, but he must drain the system when there is a danger of freezing. A system could be placed with the reflectors right next to a double-glazed skylight of the appropriate shape, but with the fin portion of the module inside a heated sunspace. You may have to experiment to find the optimum amount of insulation to put on the back of the system, since you are relying on heat from the sunspace to keep the fin from freezing. No back insulation would reduce the efficiency of hot water production and increase the night heat loss, but would certainly keep the system from freezing.

As the concept of the photovoltaic/thermal hybrid becomes better known, we will see more and more creative ways of incorporating the system into homes and other structures.

A practical application . . . .

PV/wind system in Holland.
[Photo: ARCO Solar, Inc.]

CHAPTER 13
COMBINING PV & OTHER ELECTRIC SYSTEMS

Many who now use solar electricity previously used other alternative energy systems to generate 12-volt DC electricity. These systems can be combined very easily. Figure 13.1 shows a microutility control panel developed by Skyheat to furnish the electrical power for a number of exhibits at the Real Alternative Energy Fair at Governor's State University near Chicago in 1980. The diagram on the panel was made of copper strips and actually carried the current produced by solar cells, wind generators, an alcohol-fueled generator and other devices. The current was stored in a central battery bank. The circuit used was quite simple and worked well.

Less elaborate combinations are possible for home use. Many homesteaders, when they first move to a locality without conventional electricity, use their automobile battery and alternator to furnish small amounts of 12-volt power. Kits are available to put a second (deep-cycle) battery in a car to do just this. You plug the car into the house in the evening and in the morning, while driving to work, you recharge this battery. The kits include a blocking diode so that your regular car battery isn't affected. It is possible to install a solar cell array and manually connect it to the house

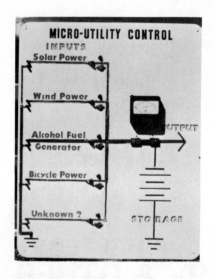

FIGURE 13.1—Microutility panel.
[Skyheat photo]

supply battery for those times when you aren't driving anywhere. (Starting a car's engine just to charge a battery is a great waste of energy and should be avoided.) The system may grow from these simple beginnings to include more solar cells or a wind generator. (Steve Cook relates his experiences with a solar/wind combination later in this chapter.)

Many homesteaders use small gasoline generators to recharge batteries or run power tools. Sometimes, when a battery bank is deeply discharged, the heavy charging current from one of these noisy devices is the only way to restore usefulness to older batteries. It is also a good idea to charge a set of batteries for an hour or so to the point where hydrogen bubbles are being generated liberally. This equalization charge allows all the cells to reach the same state of complete charge, but can usually be accomplished by putting only 10 or 15 amps into a big set of batteries. You don't have to disconnect your other sources of electric power when using one of these generators; it is possible to isolate these systems using rectifiers or blocking diodes. It is a rather simple process to convert these small engines to operate on alcohol.

Some people have installed thermoelectric generators in their remote homes. These devices, based on a semiconductor version of the thermocouple, are made by a French company and are primarily intended for delivering small amounts of power to remote radio transmitters. Operating on propane, they are inefficient and expensive to run, but if they could be fed biogas, they would be a simple, easily controlled source of low voltage. A set of thermo-couples built into the firebox of a woodstove might be another way to produce thermoelectric power. Thermocouples are low-voltage, high-current devices, so hundreds are needed to produce the 15 volts required to connect to a 12-volt battery. Nonetheless, someone may want to experiment with this concept.

Small human-powered generators, resembling exercise bicycles, are also available and can be hooked up to the same microutility network. You could build your own system using bicycle parts and an automobile generator or alternator, but a generator specially designed for this kind of use would be much more efficient.

Figure 13.2 shows the circuit we used at the Real Energy Fair. If a separate blocking diode is used as shown for each different

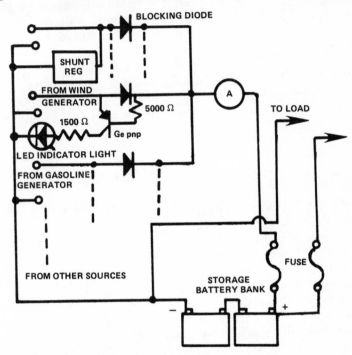

FIGURE 13.2—This circuit diagram shows several sources of alternative low-voltage electricity feeding into a single battery bank. The blocking diodes are the heart of the system, isolating the power source until its output voltage is higher than the common system voltage. The optional indicator lights comes on whenever the source is delivering power. Notice that the shunt voltage regulator is connected to the source side of the blocking diode.

source of electricity, the sources will interact very little with each other. Just make sure the diode is big enough to handle the maximum current put out by the device. However, there are some possible complications: The heavy charging current produced by an engine-driven generator could raise the voltage at the battery terminals enough to keep the solar cell array from furnishing any power while the engine is charging. Also, the voltage regulator for the wind generator may interact with the solar cell array, cutting off the wind generator when the batteries are not fully charged. You will have to experiment a little with the voltage regulators to minimize these effects.

As Steve's section points out, a properly sized wind/solar combination can furnish an almost constant supply of electricity. Adding other sources can assure continual electric power. Your imagination is the only limit in combining sources. How about water power when the creeks are up and the sun isn't?

* * *

Stephen Cook, author of the following section, is director of the North Arkansas Community College Energy Center, Pioneer Ridge, Harrison, Arkansas 72601. He is also president of CompuSOLAR, a solar computer software and educational firm.

COMBINING SOLAR AND WIND ELECTRICITY

If you are depending on a wind generator for electricity, summer breezes really do make you feel fine. That's because in many places the wind doesn't blow nearly as much then as at other times during the year. Back in 1979, I was pondering the relative lack of summer wind here in the Ozarks. I wondered how we would meet our summertime electrical needs with the 200-watt Winco wind machine we were planning to purchase. With costly excess battery storage? With a noisy, smelly gasoline backup unit? Or how? Well, we managed by combining wind and solar electricity.

Besides the general observation that solar outputs will be high when wind outputs are low (summer) and vice versa in winter, there is another advantage. Wind generators, with their moving parts, need occasional maintenance and may be down for minor (or major!) repairs once in a while. With the photovoltaic backup, these intervals are less painful. Having both energy sources makes life easier for the batteries, too.

Let's look at how wind and solar inputs into a battery bank can be balanced by first considering three separate systems.

System 1

Certainly batteries will last much longer if there is no great seasonal variation in their state of charge. Photovoltaics companies know this when they size systems for remote, constant-load applications (say for telecommunications equipment). ARCO Solar, for example, typically recommends a panel tilt of latitude plus 15°, latitude plus 20° or more (a tilt equal to the latitude will maximize fixed-panel annual production). In other words, by pointing toward the winter sun, winter outputs will be increased, summer outputs cut and a more uniform seasonal delivery of electricity achieved. By doing this, variations in the monthly kWh production can be kept within 10 to 15% of an average value— good for batteries, but wasteful of potentially available electricity.

System 2

Wind generators are hard on batteries. Consider a 200-watt Winco's output on a small tower at a good Ozark site (see Table 13.1). While 26 kWh/month is the average output, in July there is only 13.2 kWh available, yet 34.8 kWh is produced in April. The monthly production departs from the average value by nearly 50%. Without (power-wasting) voltage regulation and deep-cycle batteries, a constant remote load will be hard to power for long! And, of course, there's not much summer production.

Table 13.1—Wind Output

Typical Month	January	April	July	October	Average
Average wind speed at generator (mph)	11.1	12.0	8.5	10.0	10.5
kWh per month	30.4	34.8	13.2	23.4	26.0

How can we improve on these two situations? Consider solar first.

System 3

Suppose we want to maximize PV outputs without going to the additional complexity of tracking. By changing the tilt angle every so often, we can increase power output significantly for just minutes of work each year (if we mount our panels with this in mind). Consider a 200-watt PV array in the Ozarks (maybe six ARCO Solar ASI 16-2000 panels). Suppose we adjust the tilt four times yearly as follows:

Month	Tilt Angle
November, December, January	latitude plus 15°
February, March, April	latitude
May, June, July	latitude minus 15°
August, September, October	latitude

The average production is 9% greater than had we kept the panels tilted at the latitude angle (here 37°), which gave greater production than the latitude plus 20° case! However, the seasonal variation is greater: about 25% departure from the average value with December–January low and July high. Certainly floating-type batteries will not last as long as they would in situation 1.

Table 13.2—PV Output

Typical Month	January	April	July	October	Average
kWh per month	18.7	23.3	29.1	24.5	23.7

Based on sunshine statistics for Springfield, Missouri.

Our next situation to consider should be clear: combining System 2 and System 3 and allowing wind and solar to charge the same battery bank. Summing from Tables 13.1 and 13.2, see Table 13.3. Our biggest variation from the average is now 17%— down from 50% wind only, 25% adjustable PV only—and almost as good as the throw-away energy PV situation of System 1.

Table 13.3—Wind and PV Output

Typical Month	January	April	July	October	Average
kWh per month (wind)	30.4	34.8	13.2	23.4	26.0
kWh per month (PV)	18.7	23.3	29.1	24.5	23.7
kWh per month (total)	49.1	58.2	42.3	47.9	49.7

Based on wind and sunshine statistics for Springfield, Missouri.

In practice, 12-volt homes will not have constant year-round loads although I suspect some will roughly approach it. Certainly installations where 12-volt refrigeration is employed (as in my own) might be expected to show summer peaks, while smaller "for lights, mostly" systems might show winter peaks. If one has the proper understanding of seasonal variation of the photovoltaic (or wind) energy resource available, appliance use might be tailored to more closely match production. Of course, the kWh/month figures presented here are before losses due to transmission, voltage regulation, battery inefficiencies, inverters and obstructions which block sunlight or hinder free wind flow.

FIGURE 13.3—Hybrid system—wind/PV—constructed by Robert Sardinsky of Cataumet, Massachusetts. [Photo: Robert Sardinsky, The Sunbrero Works]

[If you know the average solar insolation $\overline{H}_T$ on a tilted surface of interest for your location in BTU/ft²-day, I've found that the relationship

$$E = \frac{\text{days in month} \times \overline{H}_T \times \text{watts of PV capacity}}{425,700}$$

will give a good approximation of the maximum kWh/month output, E, to expect from a photovoltaic array. This is a simplified relationship. More detailed assessments need to take into account array efficiency variation with seasonal temperature changes. Refer to any good wind power text for assistance in calculating the amount of electricity a particular wind generator would produce at your location.]

Regarding the hook-up of combined wind and solar systems, the following points need to be made:

1. Use blocking diodes on both positive wind and solar legs not only to prevent battery discharge but to make sure the solar cells don't motor the wind generator and vice versa.
2. Wind-only systems really require deep-cycle batteries. However, if the generators are properly sized to minimize seasonal charging variation for a location, and if the (consumption) discharge pattern is (seasonally) nearly constant, floating-type batteries could be successfully employed.
3. Separate battery banks for wind and solar could be used, with the load being powered by one bank, while the other is kept (charging) in reserve. Of course, this will be more costly.
4. I've had some trouble with my Winco voltage regulator sensing the PV generator voltage (14 volts instead of the "true" battery voltage of about 12.5 volts) and cutting out the wind output. Ideally, voltage regulators should sense battery voltage. Keep in mind that these devices waste energy and are not needed if battery capacity is adequate or you keep an eye on battery voltage and regulate appliance use as necessary. A hot water tank or a spring cleanup vacuuming job are better places to dump excess electricity than the heat sink of a voltage regulator.

Combining photovoltaics and wind is a strategy that will make sense right now for some. I suspect that in the future, as PV prices drop, families now considering wind/battery systems will be putting in wind/photovoltaic/battery systems. And, who knows, maybe even utility companies will someday see the advantages of utilizing both wind and solar in their grids.

* * *

How Does a PV Home Power System Compare to a Windcharger? Three Opinions

First of all, you must understand that when comparing apples to oranges there's nothing to say but that they are both round fruit. Ask any wind power person about the thrill of climbing the tower, the beauty of the blades turning, the massive amounts of current produced at times, and you get one view. Ask a solar electric user and you will hear about no moving parts to wear out and how the cells just sit there and magically make electricity out of sunlight.

We asked three PV Network PV/wind users what they thought.

Michael Gibbs gave some figures assuming an 8-mph wind area. Michael uses ten ARCO Solar modules acquired in a bulk purchase. He also assumes identical battery prices, home-built PV regulator and fall 1980 delivered prices.

Northwind HR^2		10 ARCO PV modules	
(2200 watts)		(33 watts each)	
HR^2	$ 7,900	Modules	$3,150
80-ft Rohn tower	$ 2,700	Regulator	50
Concrete	400		
	$11,000		$3,200

A 20-year system life is given for both the PV and the windcharger. The windcharger's estimated 150 kWh/month for 20 years equals 36,000 kWh. The PV array average estimate of 40 kWh/month for 20 years equals 9,600 kWh. Windcharger: 30.5¢/kWh; PV array: 33.3¢/kWh. His conclusion is that the windcharger is a better investment if wind speeds average about 10 mph, but that PV is competitive in low-wind areas.

Steve Willey, a wind enthusiast who uses two Winchargers and a PV array, agrees that PV and wind are about the same in watts per dollar but points out that distribution seasonally and week-to-week is the main difference. He thinks that homemade towers can keep costs down with wind systems, but wiring from the distant windcharger can hike the costs up again. He recommends PV in all cases, instead of a second wind machine.

Val Bertoia, an artist who uses wind power, PV and hydro, and who builds wind power systems for sale, says instead of comparing the two for competing prices, simply use both power sources of moderate quantity.

A practical application . . . .

South view of a PV-powered, stationary Airstream in New England.
[Courtesy: Jay Baldwin; Photo: Robert Sardinsky, The Sunbrero Works]

CHAPTER 14
MARINE & MOBILE
SYSTEMS

Not all homes are fastened to the ground. Some people choose to live aboard boats; others live in recreational vehicles or camping trailers. For these nomads, a self-sufficient source of electrical power becomes a necessity. Otherwise, there is the constant problem of finding a camping area or boat dock with utilities, with the lack of flexibility and costs that this entails. Gasoline or diesel generators are noisy, need constant attention, are expensive to repair, and have an appetite for petroleum products. There is nothing that can destroy the mood of an idyllic spot faster than a running generator bringing its bit of city pollution to the wilderness. You, of course, can and should be able to charge your batteries from your engine when traveling from place to place, but starting the engine just to furnish electric power is very wasteful and hard on the motor. This problem is especially acute on sailboats, where most travel is without the use of a motor. Even for the casual traveler or boater who just wants to get away for the weekend once in a while, a small but dependable source of electricity is important.

The solar cell is the perfect answer to the nomad's electrical needs. The arrays are rugged, dependable, quiet, low-maintenance and relatively unobtrusive. All the reasons for using photovoltaics in a fixed remote home become especially valid for the camper or boater. The principles of sizing, installing and using the systems are similar, and the information given in the rest of this book does apply to marine and mobile use. However, there are some differences and special problems, and in this chapter we discuss how to tailor photovoltaic systems to these needs.

MARINE USE

The marine environment is especially harsh. Equipment that will last forever in a house falls apart or hopelessly corrodes in a few weeks on a boat, so special care must be taken.

On board, photovoltaic electricity becomes more than a convenience. A sailor must provide for all energy needs while cruising. The ability to charge a dead battery at sea is a safety requirement important enough that, soon, all ocean racers will be required to install photovoltaic systems. While some sailors choose to simplify their power requirements by using kerosene cabin and running lights, the wide range of electronic and electrical devices now available makes some low-voltage power source desirable, if not absolutely necessary. (Kerosene is becoming expensive and hard to find, and seems to have deteriorated in quality.)

MARINE SOLAR CELL PANELS

Most cruising boats are wired for 12-volt DC and come equipped with deep-cycle storage batteries, so adding PV panels is relatively simple. The solar cell module must be specially designed for use in a marine environment, however. In the past, some distributors have sold conventional modules to boat owners with disastrous results. The metal case frames were quickly eaten away by salt water, water seeped in and corroded the internal wire leads, or the glass cover plates cracked, leading to failure. The first Skyheat marine panel, made in 1975, lasted only two years on a sailboat before succumbing to corrosion and cell breakage from vibration and jarring.

Several manufacturers now make well-designed marine solar battery chargers. These include Free Energy Systems, PDC Labs, Solarex and SunWatt (addresses are listed in Appendix G). PV

FIGURE 14.1—A neat installation of a marine PV module in a tight location.
[Photo: Free Energy Systems, Inc.]

modules for marine use should have a waterproof acrylic or poly-carbonate cover and be sturdy enough to walk on. An embossed, nonskid surface is more efficient in gathering light coming from an angle. The wires or terminals are extra-strong and corrosion-resistant. The sizes available range from 5 to 18 watts output. These units are smaller than the standardized modules for the solar electric home, so more would have to be used in parallel to get the same power capacity. Unfortunately, marine modules are also more expensive per watt. Some expense is added by the more rugged design, but much of the high cost is created by the small market and complex marine distribution system. If enough boat owners demanded these systems, some manufacturer would most likely deal directly in a high-volume fashion, causing a drop in prices. The June 1982 issue of *Cruising World* has an excellent set of articles about solar and other alternative power sources.

HOW BIG A MODULE

Marine modules are sized in the same way as residential modules and the marine user can refer to Chapter 4 on sizing as a guide. However, there are two complicating factors that must be considered: the amount of time the boat is used, and the fact that when the boat is in motion, it faces many different directions, so that a fixed mount tilted south is not always appropriate.

Let's consider the use of the boat first. Some small sailboats get only casual weekend use and spend the rest of the time tied to a dock or mooring. The chief reason for having a solar charger is to keep the marine battery trickle charged without the hassle of periodically hauling it home to charge or the additional expense of renting dock space with utilities. Under these circumstances, the solar panel is sized to the battery to give a sufficient current to keep the battery charged without damaging it. A 5- to 7-watt PV panel will take care of a 100-ampere-hour marine battery. No voltage regulator will be necessary in this situation, making the installation very simple. For large systems on bigger boats used in this manner, simply use the ratio: 5 to 7 watts for each 100 ampere-hours of battery capacity. Too small an array won't produce a sufficient trickle charge to the battery set, and sulfation could permanently damage it. A large array will do no harm but might require a voltage regulator unless the solar module is designed to be self-regulating, as are the SunWatt and Solarex modules.

For the boat that gets more regular use, extended two-week cruises interspersed with regular overnight sails, the module size could be doubled. Since solar arrays are expandable, you can add extra modules as your needs change. One way to judge your electrical needs without complex calculations is to use what you consider the minimum system for a period of time. If you find that you are constantly watching your electrical usage and still have dead batteries, add another panel.

For those who are into extended cruises or live aboard their boats, both the solar cell array and battery size will be determined by electricity usage. Again, refer to Chapter 4 to calculate your needs. The list on page 46 gives the current draw for many of the appliances you are likely to operate. The current needed by some marine devices is given below:

Bilge Pump—3 amps
Depth-Sounder—0.4 amp
UHF-Receive—0.5 amp
UHF-Transmit—1 amp
Running Lights—0.5 to 1.5 amps

You can often find the current draw or wattage of a marine electronic device by looking on the name tag, or you can measure it with a multimeter. Multiply the current in amperes by the hours of expected use per day to obtain the ampere-hour requirements of the device.

If your list of desired appliances and devices is extensive, and the size of your solar array is getting out of hand, there are two things that can be done: (1) Cut back on your electricity usage. Do you really want to cart all those toys around with you when you're getting away from it all? And how many hours per day or week will you really need to run them? (2) Increase your power-generating capacity by adding a windcharger or water-driven generator. If blocking diodes are used, all these devices as well as the generator or the auxiliary engine can feed the same battery set. (Chapter 13 details how to do this.) Even with reasonable power needs, the battery set required to ensure seven days' storage would be far larger and heavier than you would want to carry. The sizing calculations were made for people with stationary installations and lots of room, and you could probably get by with half that battery capacity. If you have four sunless days, power down unless your wind generator is putting out a good charge.

MOUNTING AND WIRING

Mounting a solar panel on a boat involves compromise. There are two basic schools of thought on mounting:

1. Mount the device permanently, flat on the deck or cabin top or some other spot that isn't likely to get shaded and hope that, on the average, you will receive a significant portion of the available sunlight. This method works best in the summer or near the equator where the sun is high overhead. For the casual user who moors on a buoy, the system has the advantage of working independently of the way the boat points as it swings around. An added advan-

tage is that the permanent mount can be fixed with through-bolts so that the module is difficult to steal.

2. Don't permanently mount the device, but arrange it so that it can be tilted in the direction that produces the best operation at the moment. This method requires considerably more attention but can probably double your output —if you are reasonably conscientious about adjustment. If the boat is usually docked in one spot, the mount can be semipermanent, as it can be on a long passage in one direction.

Remember when selecting a mounting location that shadowing one cell essentially turns off the entire series string. Acrylic plastic and polycarbonate both expand and contract a great deal more than wood or fiberglass with temperature changes, so the mounting screws should be considerably smaller than the holes in the module. Use polysulfone rubber caulk to seal the holes around the screws and the points at which the connecting wires go through the deck. An elegant way to mount a solar module is to build a wood frame (preferably teak) and place it around the panel, sealing the groove with polysulfone caulk. Fasten the frame to the deck but not the panel, allowing it to "float" and move with temperature changes. The edges and wiring terminals, the spots most likely to develop leaks, will be protected by the frame.

Most marine solar modules have built-in blocking diodes, which makes wiring very simple. Just run two wires from the module to the battery, connect the positive wire from the module to the positive battery terminal, and likewise for the negative. Small amounts of current can easily be carried by No. 18 wire if the distance to the battery is less than 20 feet; above that, use No. 16 wire. Use No. 16 also if the expected solar output reaches about 1.5 amps. For larger installations, refer to Chapter 10 on wiring. An ammeter installed in the positive wire from the panel will tell you how well your system is working, and a fuse placed as close to the battery as possible is a safety must. The fuse will only carry the current from the solar cells and should be small.

For multiple battery systems, there are several wiring options:

1. The solar module can be wired into the main electrical buss at the battery select switch so the module is switched from one battery to the other with the main charging system. This has the advantage of simplicity but will leave one or more batteries without charging at any one time. You

would have to periodically switch to each battery to ensure uniform operation.

2. Use a separate module for each battery, wired directly to the batteries. This is the best system, but expensive as two smaller arrays cost more than the equivalent wattage in one large array.

3. If you can obtain a large output marine array without a built-in blocking diode, run a separate wire to each battery from the single positive and negative terminals of the module and install a separate blocking diode in each positive wire. In this system, the battery with the lowest state of charge will take the lion's share of the current but, eventually, all the batteries will be equally charged.

Once you begin to enjoy the freedom of solar electricity, you might want to think about other ways the sun can enhance your boating comfort. Pat Rand Rose, author of *The Solar Boat Book*, offers instructions for making solar hot water heaters and other specifically marine devices.

MOBILE LAND USERS

The nomads who stay on solid ground share needs and problems with boaters, and many of the suggestions in the marine section apply equally well to them. This section offers information oriented toward using solar cells on recreational vehicles and travel trailers.

There is a developing network of enthusiastic PV users who, through experience, have solved many problems related to adapting PV modules to moving vehicles. People like Noel Kirkby and Phred Tinseth have become good information sources and much of the following is based on their knowledge and input.

SELECTING AND SIZING THE SYSTEM

Since RV users rarely encounter the corrosive salt water atmosphere that plagues marine environments, more conventional modules can be used. This means that the same amount of money will buy a much larger PV system. Even though there is extra power available, you must still be conservative and flexible in your power requirements. Low-voltage quartz-halogen and fluorescent

FIGURE 14.2—A photovoltaic starter kit for the recreational vehicle user, which includes a simple tilting mount. Most RVers will probably need a larger system and a larger mount with several adjustment positions.

[Photo: Noel Kirkby, Solar Electric Systems]

lights are more efficient than, and preferable to, cheap automotive light fixtures or 12-volt versions of conventional incandescent lights found in factory-wired motor homes. The usual 12-volt RV refrigerator is an energy hog with insufficient insulation, and will gobble up the output of two or three 35-watt PV modules. (See Appendix D for the latest solutions to this very important problem.) Ordinary air conditioners are out, but other ways to keep cool will be discussed later. You can buy 12-volt versions of many kitchen appliances like electric frying pans, blenders, and even popcorn poppers, but these devices draw an enormous current and many now available are shoddy and not worth having.

Once you have determined your appropriate electricity usage, refer to the sizing information in Chapter 4 to calculate the solar cell wattage you will need for the locations you expect to visit— one to three 35-watt PV modules for the average user. Storing electricity in batteries is more a problem in motor homes than it is at a fixed home or on a large cruising boat, where the extra weight isn't so important. Planning for a week's worth of storage is out of the question, but there are compensating factors in the ability to change the batteries from the main vehicle motor.

MOUNTING AND WIRING THE SYSTEM

Mounting the solar cell array on an RV is much simpler than on a boat. It nonetheless requires considerable thought and care. There are three mounting options:

Fixed Mount

If you travel frequently, mount the module flat on the roof; if the RV is always parked in the same spot, tilt it to the appropriate angle. In northern latitudes, very little winter sun will be picked up by a horizontal array, but the performance will improve in the summer. For the weekend camper, this might be the best and most trouble-free system.

FIGURE 14.3—A good example of mounting solar cell modules permanently in a flat position on a motor home. Note the small air space under the mount to keep the cells cooler.   [Photo: Noel Kirkby, Solar Electric Systems]

Tilt-up Mount

One great advantage of a motor home is that the solar array will be needed only when the vehicle is parked. A mounting device that lies flat during travel, then tilts up and, possibly, swivels to face south, is a great improvement over the fixed mount. It adds another chore when setting up or breaking camp, but some of the ingenious mounts built onto motor homes or trailer roofs make the job quick and easy.

Separate Ground Mount

People like to park under trees, creating a pleasant shady environment that compensates for the meager insulation present in all too many recreational vehicles. But unless you jockey the position just right to pick up the sun, your roof-mounted solar array will produce no power. The solution is to build a rack to hold the solar cells that can be put out in the sun. Keep the cord length under 50 feet or the voltage drop in the lines will significantly lower your output. Use No. 14 wire for one or two 35-watt panels, No. 12 for three or four. These wire sizes will produce a 1-volt drop in the wire under the worst circumstances of distance and sunlight intensity. If in doubt, consult the wire table and chart (page 98) to determine your needs.

However you mount the system, make it secure. The self-produced 55-mph winds as well as the thunderstorms and gusts encountered while parked will exert tremendous force on the mounts and fasteners. Modules are expensive and don't take a lot of flexing. Most manufacturers suggest proper mounting procedures.

Since recreational vehicles are already wired for 12 volts DC, adding a PV system would seem to be a simple wiring job. It is, but the job of redesigning and upgrading shoddy factory-installed wiring may get complex and expensive. Unlike most marine systems, which are traditionally wired with sturdy components neatly laid out, many RVs have a handful of poor-quality automotive components and wire too small to carry any decent load. Take a look at Chapter 10 and compare the recommendations to your situation.

A well-designed photovoltaic system installed properly will give you the independence you desire with little trouble and maintenance. Keep the faces of the modules clean and take care of your batteries, and the system will furnish continuous power for many years.

Skyheat is working on a hybrid PV/hot water collector that will be applicable to mobile use. Aluminum fins covered with solar cells, the module should be flat enough to be unobtrusively mounted on a boat or RV. This will allow the luxury of hot running water without using expensive LP gas. Other possible uses of solar energy include passive heating and even solar-powered air conditioning in a compact package. In the future, we should see more exciting ways to achieve an independent, flexible style of nomadic living.

* * *

Phred Tinseth has written some thoughts and suggestions on batteries, appliances and PV/RV living that are included below.

Batteries

Everything written on the subject to date points out the need for deep-cycle batteries of large capacity. There are exceptions in the RV field. An example is the RVer who is on the road much of the time. Between the vehicle's charging system and panel(s), the battery is almost constantly fully charged—and frequently overcharged. Automotive batteries will do in some cases, but not if the vehicle is parked for extended periods. A large battery bank, weighing hundreds of pounds, can be accommodated by a fixed-property owner. An RVer can afford to haul around 200- to 400-ampere-hour capacity only so long before the effort becomes counterproductive. Overcharging, rarely of concern in a fixed installation, can easily occur in an RV if it is parked without being used. (This should be obvious, but people don't always consider the obvious.) In an RV, overcharging can occur even when it is being lived in fulltime. Refrigeration, water heating and cooking, for instance, are usually done with LP gas. Cooling fans are not always in use. Similarly, lights are not used much during certain seasons. The RVer is outdoors most of the day. A few hours of TV and small lamps are used in the evening. In late spring, part of the summer and early fall, one ARCO Solar or equivalent panel can easily provide over 20 ampere-hours daily to a battery that is required to provide only 5 ampere-hours daily. The use of switches, meters to monitor charging, regulators, or self-regulating solar modules can solve this problem.

What to do with surplus? Shunt regulators are available. In a residential situation they are frequently hooked up to a stainless-steel strip, lamp or more batteries. In an RV, where the water heater is of only 6 to 10 gallons capacity, the excess can easily be shunted to a heating element. I have seen advertised a standard-size 12-volt DC element. It is not difficult, though, to tap a water heater and insert a soldering iron element or one of those auto-mobile immersion-type elements. Also, one can use more recharge-able batteries for some tasks and Mike Hackleman's "black box" charger to take up the excess.

Air Conditioners

Surely you've seen the Coleman and other brands that sit on the roofs of most RVs. They all run off an inverter of about 1600 watts capacity. The problem is that to run them for any length of time requires a battery bank heavy enough to flatten tires. If you're a property owner, you're stuck. If you're a fulltime RVer, you should move to the cool country. In a humid climate you can often cool the place nicely with a 12-volt low-amp-draw fan. If you're in the western low-humidity country, you can run an evaporative (swamp) cooler. Selection of a fan motor, for instance, is critical. A typical 12-volt motor to run a fan might easily draw 10 amps. This is unacceptable. On the other hand, a so-called solar motor of 1-amp draw will not move enough air. But, a high-torque DC permanent magnet motor of about 2200 rpm no-load, rated for 27 volts DC will move plenty of air at 12 volts and will do so at between 2.5 and 3 amps. (Incidentally, Noel Kirkby sells one for $29.50, stock number M-8). Couple this with the Herbach and Rademan 12-volt pump and you've got effective cooling at less than 3 amps. [Note: This pump has no check valve. Rather than prime each time, you could install a check, but why go to the expense? Rather, put a loop in the tubing to the upper water distributor. Enough water will stay in the line to allow on/off without priming.]

As I write this, it's 105° here in the Arizona desert. My cooler is keeping my 30-foot RV at 82°. Not bad at a little less than

3 amps. The cooler body? One of the typical, roof-mounted mobile home boxes. Not very sleek, but so what? You can get old ones for nothing. They don't interfere with wind on the road if you drive 55 mph (50 is 20% cheaper). Several manufacturers make aerodynamic evap coolers for RVs. They cost up to $400+ and use over 6 amps in most cases. If you want a slick-looking rig, buy one. But throw the fan motor out and get a good one.

More on fan motors. Per the electrical laws, if you run a 20+-volt motor at 12 volts, amp draw will remain the same. True, except that the laws fail to mention that the fan blades are not having to "push" so hard at air resistance, so amp draw will be reduced—significantly.

Tools

Drills are the easiest tools to convert. Get a rechargeable Skill 3/8-inch model and either recharge it from a Hackleman black box or, better yet, remove the batteries and wire it directly to 12 volts. The drill motor on the new ones is 6 volts, but runs better on 12, unless it's run too long. Mine now has a switch and dropping resistor for 6- or 12-volt operation. My 1/2-inch drill is a standard 110-volt model. Remove the motor and replace it with one that matches in size and has the necessary torque (find one in the Pittman catalog).

Nobody in an RV has to carry about a large number of big power tools, but I do just for fun. They are not converted to 12-volt DC. The only reason I mention this is to illustrate that there are times when an inverter can be useful, though I still don't need one. The property owner does, but the RVer can stop for a few days where there is 110-volt access, make the necessary repairs or whatever, and move on.

It is easy to convert belt-driven tools. Many RVers are rock and gem hunters. The machines they use in this hobby/profession are almost always belt-driven and can be converted, literally, in minutes to low-voltage if the proper motor is selected. Grain grinders and sewing machines—the same.

Blenders/Food Processors

Take out the motor. Find one that will fit inside the casing and that has a high torque. Don't worry about amp draw. I've never needed to run a blender more than a minute or two. Amp draw is insignificant.

Washing Machines and Dryers

The old-fashioned washers work very well and are easily converted to 12-volt DC. Dryers are not to be used—neither in RVs nor homes—except the passive solar variety. In an RV, though, there is not the water or space for an old-fashioned washer. If people are on the road, it's an easy matter to toss dirty clothes into a covered container and allow vehicular motion to do the scrubbing. (A full day on the road will do a much better job than an ordinary washing machine.) At day's end rinse and hang to dry. Or, better yet, stop and use a laundromat once in a while— though laundromats are extraordinarily depressing places. I do know one RVer, with a very large rig and a 24-volt system, who has a washer/dryer combination. He also carries 100+ gallons of water at 8+ pounds per gallon, and has over 40 feet of space. This is clearly not the norm.

APPENDIXES

A SHORT COURSE IN ELECTRICITY

Electricity is electromotive force (voltage) and electron flow (current). If we think of a wire or conductor of electricity as a water hose, voltage would be the water pressure and current would be the rate of flow. Thus, 40 pounds per square inch at 4 gallons per minute is similar to 12 volts at 2 amperes. To expand this analogy, we can compare the size of the hose with the size of a wire. A large hose will let water flow with less restriction; a large wire will let electricity flow with less resistance.

We use two types of electricity: alternating current (AC) and direct current (DC). Our most common use of DC is in flashlights, portable radios, and cars and trucks. All of these devices are energized by the direct current from batteries. Most newer stereos and TVs use direct current but, with transformers and rectifiers, first change the 120-volt AC house current from the wall outlet into low-power direct current. Direct current is a one-way flow of electrons from minus to plus.

In our homes we normally use alternating current—the flow of electrons first in one direction and, then, reversed in the other direction. Typical house current is between 110 and 130 volts and alternates or cycles 60 times each second. The frequency of ordinary house current is therefore 60 hertz (Hz).

Alternating current is used almost universally because it can be transmitted over long distances of wire at high voltage with little power loss. Low-voltage DC wire loss is significant. That's why automobile jumper cables are short and thick.

At the present time, large-scale production of alternating current is more economical, although new developments in solid-state technology are making high-voltage DC transmission systems attractive. Small-scale DC power production can be economical, too. Years ago, many rural residences produced their own DC electricity. These home power plants were fueled by gasoline, diesel fuel, or the wind. Several firms manufactured appliances which could be powered by these home generation DC plants. With the spread of alternating current through the government's Rural Electrification Administration's programs, the nation's thousands of home power plants fell to disuse and were replaced by subsidized "cheap" alternating current from centralized large power plants.

With rising electric rates, more and more people are returning to home-grown DC power plants. In addition, new technology and the recreational vehicle boom have helped the small-scale electricity producer by providing a wide range of tools, equipment and appliances which use DC while consuming very little current (compared to those used in the 1930s). More efficient motors, DC fluorescent lamps, solid-state TVs, stereos and radios are just some of the devices now available to the small-scale energy producer. New solid-state inverters which convert 12-volt DC to 120-volt AC make it possible for us to use all the appliances we have become accustomed to using with house current.

VOLTS, AMPS AND OHMS

Many people have had just enough exposure to electricity in school to confuse and frighten them. The information you need to design and wire your own solar electric home is really quite simple. First the terms:

Voltage (E), measured in volts, indicates the electrical "pressure" in a system. The output of a flashlight cell is only 1.5 volts DC while the typical automobile battery puts out around 13 volts (even though it is called a 12-volt battery). The symbol for volts is V.

Current (I), measured in amperes (amps), describes how much electricity is flowing in a circuit. The symbol for amps is A. The current flowing through the starter when you start your car can be 200 amps or more, while that drawn by a portable radio may be only 0.1 amp or 100 milliamps (one milliamp, symbolized mA, is one-thousandth of an amp).

Resistance (R), measured in ohms, is just that—the resistance to the flow of electricity. It is a property of the electrical device or appliance. Sometimes the word **impedance** will be used instead of resistance. Impedance is a complex form of AC resistance that can change with the frequency. For a DC system, we will concern ourselves only with the resistance and not worry about the complex impedance of, for example, a fan motor. If a fan is designed to operate on 12 volts, and we are using 12 volts, we are concerned only about the current draw when the fan is operating properly. The symbol for ohms is Ω.

Power (W), measured in watts, indicates the rate at which electrical energy is being used. The power in watts is the voltage multiplied by the current used: 1 V x 1 A = 1 W (1 volt x 1 amp equals 1 watt of power). W is the symbol for both watts and power in general. Horsepower is the English measure for power; 746 watts is 1 horsepower. Sometimes you will see power stated in kilowatts (kW), a kilowatt being 1000 watts.

Energy is measured, for our purposes, in watt-hours (Whr) or kilowatt-hours (kWh). If you run a 25-watt light bulb for 1 hour, you have used 25 watt-hours of electrical energy. If you left that bulb on for 100 hours, the total energy used would be 25 x 100 = 2500 Whr or 2.5 kWh. It is very important not to confuse energy and power. Power is the rate at which energy is used or changed.

Ohm's law is the relationship between volts, amps and ohms. This law can be explained by Figure A.1. To make the wheel chart more complete and useful, power (W) is also included.

Don't let the chart scare you. It's really very simple and graphically shows the relationship between volts, amperes, ohms and watts. We have already seen how volts and amperes are related and how resistance (ohms) can affect current flow. Watts is just another name for the product of volts times amperes.

This chart is a useful tool because we can refer to it to change values from watts to amperes as we size our PV power systems. We can also use it to determine the size wire we need to minimize resistance. Here are a few examples of the chart's use:

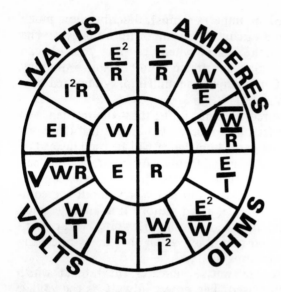

FIGURE A.1—Ohm's law.

Example 1. Fluorescent Lamp

Let's say we want to power a fluorescent lamp with a PV array. The lamp is 12-volt DC rated at 20 watts. Our PV module is rated at 35 peak watts. That seems to mean that the lamp can be powered by the PV module with some power to spare.

The same module may actually produce only 12 volts at 2 amps output. Thus, we find that the rated output (35 peak watts) is really the maximum output under perfect conditions. By multiplying volts by amperes, we find that the usable wattage is really 24 watts—still enough to power our lamp.

But we want to use the lamp at night, having produced power during the day and stored it in a 12-volt DC battery. If the sun usually shines 6 hours per day where you live, your PV module will produce 6 hours times 2 amperes or 12 ampere-hours per day. Another way to describe this is to say that the PV module output is 144 watt-hours per day (12 volts DC x 12 ampere-hours = 144 watt-hours).

If our lamp draws 20 watts, then we should be able to operate the lamp over 7 hours per night (144 watt-hours ÷ 20 watts = 7.2). However, there is some internal loss in batteries (about 30%) and some wire loss (another 5%), so figure on about 5 hours of actual operation per night for that lamp. Of course, short winter days or cloudy skies will change the production output, so PV systems are generally sized with weekly or monthly solar output values.

Use the chart to change watts to amperes and back to watts. Batteries are rated by ampere-hours of storage capacity, but most appliances are rated in watts. We now can see how easy it is to go back and forth from watts to amperes. Example: 20 watts at 12 volts DC = 1.6 amperes ($I = W/E$).

Volts divided by amperes will give you the resistance of a circuit. Resistance becomes an important factor with low-wattage (low voltage times amperage) power systems.

Example 2. Resistance Heater

Portable electric heaters are quite often used to take the chill off a small room. These inexpensive heaters typically have a wattage of 1200 watts and are designed to be used on 120 volts AC. If the heater has no fans or controls, only a resistance element, it will operate on DC just as well as on AC.

What is the current draw of this heater? From the chart, $I = W/E$. The current is the wattage divided by the voltage

$$I = \frac{1200 \text{ W}}{120 \text{ V}} = 10 \text{ A}$$

The resistance of the heater can be calculated from Ohm's law, $R = E/I$. Putting in the numbers:

$$R = \frac{120 \text{ V}}{10 \text{ A}} = 12 \ \Omega$$

So the heater wire has a resistance of 12 ohms. You could connect the heater to a 12-volt DC electrical system without any problem. What would be the current draw in this situation? Taking the resistance of 12 Ω, which is a property of the wire essentially independent of the applied voltage, we use the chart again:

$$I = \frac{E}{R} = \frac{12 \text{ V}}{12 \ \Omega} = 1 \text{ A}$$

What wattage would the heater put out? W = EI = 12 V x 1 A = 12 W. So on 12 volts, the 1200-watt heater would deliver only 12 watts—barely enough to keep warm. You could use an ordinary portable electric heater on 12 volts DC (provided you disconnect the fan, if it has one), but it will produce very little heat.

Example 3. Wire Resistance and Voltage Drop

A matter of concern when doing any kind of electrical wiring is the size wire to use. Chapter 10 examines the question in detail, but we have included an example here to show how you can determine the loss expected when a particular size wire is used for a certain job.

No. 10 copper wire has a resistance of 0.1 Ω per 100 feet. If you hook up a battery to a 100-watt, 12-volt light bulb in the manner shown in Figure A.2, the total path length is 50 feet (25 feet to the bulb and 25 feet back). What voltage will be delivered to the bulb? First, let's calculate the resistance of the wire:

$$0.1 \ \Omega \ \times \ \frac{50 \ ft}{100 \ ft} \ = 0.05 \ \Omega \ \text{wire resistance}$$

which doesn't seem like much. The resistance of the bulb when it is operating normally comes from Ohm's law and our chart (R =

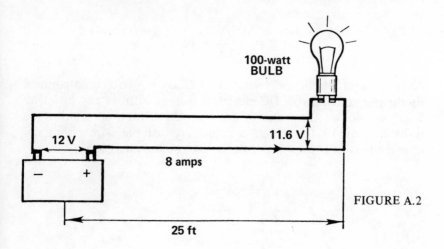

100-watt
BULB

12 V

11.6 V

8 amps

25 ft

FIGURE A.2

E^2/W is the formula we use since we know the normal voltage is 12 volts and the design wattage is 100 watts):

$$R = \frac{12^2}{100} = \frac{144}{100} = 1.44 \ \Omega$$

The total resistance of the system is the sum of the bulb plus the wire resistance, or 1.49 Ω. The total current drawn from the battery is given by the formula $I = E/R$ (notice how we keep using the same formulas over and over):

$$I = \frac{12}{1.49} = 8.05 \ A$$

which we can round off to 8 amps. The 8 amps flowing through the No. 10 wire will cause a voltage drop; the lamp will not get the full 12 volts the battery is delivering.

To calculate the voltage drop, simply use Ohm's law again: $E = IR$. The voltage drop, $E_w = 8 \ A \ x \ 0.05 \ \Omega = 0.4 \ V$, is determined by taking the current through the wire and multiplying it by the wire resistance. If 0.4 volt is lost in the wire, $12 - 0.4 = 11.6$ volts is delivered to the bulb. The important thing to notice is that the voltage drop in the wire is directly proportional to the current the wire is called upon to carry and also directly proportional to the wire resistance. The message is, if you want to carry a lot of current, you need thick, short wires.

The last calculation in this example is the power loss in the wire. From the chart, the power $W = EI = 0.4 \ V \ x \ 8 \ A = 3.2 \ W$. The power actually used by the nominal 100-watt bulb is 11.6 V x 8 A = 92.8 W. This means the total power taken from the battery is 92.8 W + 3.2 W = 96 W. Another way of arriving at the same conclusion is to multiply the total current by the total battery voltage (12 V x 8 A = 96 W). Even the No. 10 wire, in this case, reduces considerably the actual power delivered by the light bulb. The bulb is cooler and will not be as bright, so the amount of light lost will be proportionally even larger. The only good thing about this situation is that the bulb will last longer.

Using the techniques in this example, you can now calculate the voltage drop and power loss in wires of various sizes in your own situations.

SERIES AND PARALLEL CIRCUITS

Series

Series circuits are probably the easiest for the beginner to understand. The wire resistance example had three resistances in series, as shown schematically in Figure A.3. There are two wires, each with its own resistance, R_1 and R_2, and there is the resistance of the load, R_3. The current from the battery has to go through all three resistances to get back to the negative terminal of the battery. The total resistance is simply the sum of the individual resistances:

$$R_{total} = R_1 + R_2 + R_3$$

This series circuit is like old-fashioned Christmas tree lights—if one light burns out, it breaks the circuit and no current flows at all.

RESISTANCE OF WIRE TO LOAD

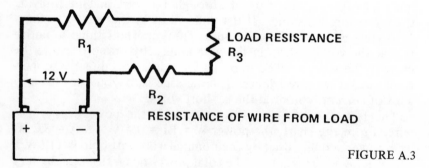

LOAD RESISTANCE
R_3

12 V

R_1

R_2

RESISTANCE OF WIRE FROM LOAD

FIGURE A.3

Parallel

A parallel circuit is easy to understand if you think about the current going through each part (Figure A.4). Each load resistance is connected directly across the battery. R_1 might be a 12-volt light bulb, R_2 a 12-volt soldering iron, R_3 a portable radio. The parallel circuit is what you end up with when you wire your house. Each load is more or less independent of the others and draws its own current as calculated by Ohm's law ($I = E/R$). The

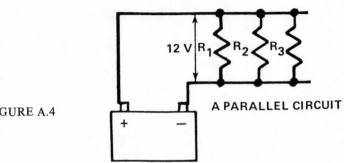

FIGURE A.4   A PARALLEL CIRCUIT

total current will simply be the sum of all the individual currents. In the parallel circuit, since each device is connected separately to the battery, turning off one device should have little influence on the others, provided that the connecting wires are large enough so that the wire resistance in series with the devices is small.

$$R_{total} = \frac{1}{\frac{1}{R_1} + \frac{1}{R_2} + \frac{1}{R_3}}$$

USING PV WITH AC

This section is included to show the expanded range of possibilities that inverters have given to DC photovoltaic installations. It also shows how an existing PV installation can be expanded to include more appliances and to utilize conventional AC equipment. It is hoped that this information will inspire some of you to plan your present PV system with the future in mind.

Note: The following information is based on an experimental system which has not yet been tested. Some components listed may not be necessary or even desirable.

DESIGN CRITERIA

The PV industry considers 12-volt DC photovoltaic systems and 120-volt AC photovoltaic systems distinct and separate, using different design and equipment approaches for each system. This distinction is based on what appears to be different types of load requirements. However, the recreational vehicle industry has bridged any gap between DC and AC systems with highly efficient small inverters and energy-efficient solid-state devices. By converting individual circuits and tasks from DC to AC or vice versa, a smooth and economical transition can be made. An example is the changeover in lighting from incandescent lights to more energy-efficient fluorescent lamps. By separating circuits requiring either "raw" conversion of direct current to alternating current (*e.g.*, refrigerator/freezer) from circuits requiring conversion and frequency control (*e.g.*, television, stereo), less costly inverters can be used.

MODIFICATION OF EXISTING PV POWERED HOME

Present PV Power System Components

- PV Array: Four each 35-watt ARCO Solar modules
- Mounting: Roof-mounted "Boltmaster" galvanized steel assembly hinged for seasonal altitude adjustment
- Regulator and Distribution Box: One each ARCO Solar Village Power Panel
- Battery Bank: Six each C&D 2-volt batteries with an 8-hour discharge rate at 52.5 amps
- Auxiliary Power Source: gasoline-powered 12-volt DC generator
- Supplementary PV Array: One each 2-amp Free Energy Systems PV array, one each 40-watt inverter, two each 6-volt Gould batteries rated 22.5 amps for 8 hours discharge
- Appliances and Equipment: The home presently has fluorescent lamps and small incandescent lights in all five rooms; black and white television, radio/stereo record player/tape recorder, fans and water pump

Modifications

- Enlarge PV array to a total of 30 each 35-watt solar cell modules
- Rebuild PV array mounting and install array at ground level
- Replace regulator to carry increased power production
- Install DC to AC inverters for individual circuits
- Install testing and monitoring equipment

The Solar Electric Home
After Modifications

Essentially, the power system is a unique interface between an energy-conserving conventionally powered (120-volt AC) home with photovoltaic hardware designed for economical (12-volt DC) production. By linking the two components with properly sized DC to AC inverters, the transition is made. Power storage for nights and cloudy days will be the existing battery bank in parallel with the PV array, thus making the demonstration home autonomous. For homes connected to the utility power grid, batteries would be replaced or supplemented by a synchronous inverter.

ELECTRICAL CONSUMPTION PROJECTIONS

Although the following figures may seem low, they are based on the experience of living for over two years in a PV-powered home. In addition, these figures are cross-correlated with those of others who live in PV- and wind-powered homes. The AC projections are based on information gathered over the past ten years from several households that have cut electrical consumption without sacrificing comfort.

Present 12-Volt DC PV System

- Lights: Two each 16-watt fluorescent lamps for 4 hours per day (3.84 kWh per month)
- Television: 16 watts for 4 hours per day (1.92 kWh per month)
- Radio/Stereo: 8 watts for 1 hour per day (0.24 kWh per month)
- Water Pump: 72 watts for 0.5 hour per day (4.32 kWh per month)
- Total: 10.32 kWh per month

Projected 120-Volt AC PV System

- Lights: Four each 16-watt fluorescent lamps for 4 hours per day (7.68 kWh per month)
- Television: 135 watts for 4 hours per day (16.2 kWh per month)
- Radio/Stereo: 105 watts for 1 hour per day (3.15 kWh per month)
- Water Pump: 335 watts for 0.5 hour per day (5 kWh per month)
- Washing Machine: 375 watts for 2 hours per week (3 kWh per month)
- Refrigerator/Freezer: Continuous duty with 330-watt draw (70 kWh per month)
- Appliances: Vacuum cleaner, clock, blender/food processor, sewing machine, fan, tools, etc. (5 kWh per month)
- Total: 110.03 kWh per month

ELECTRICAL PRODUCTION PROJECTION

Based on an existing 20-module array in an Arkansas locale: 30 each 35-watt solar cell modules will produce 990 watts at noon in full sun, or in January—99 kWh per month, in August—151.5 kWh per month, average—125.5 kWh per month.

The existing battery bank will be used. The current draw of the home is 33.342 ampere-hours per day; 400 ampere-hours of battery storage will provide electricity for approximately 8 days at 70% battery efficiency.

CIRCUIT SCHEDULE

- Lights: Eight each 16-watt fluorescent lamps with 200-watt inverter in circuit
- Television and Radio/Stereo: 500-watt regulated frequency-controlled inverter in circuit at junction box
- Water Pump: 500-watt inverter at junction box
- Refrigerator/Freezer: 500-watt inverter at junction box
- Washing Machine and Appliances: 500-watt inverter at junction box
- Additional Notes

1. "Power Miser" (watt saver) at washing machine, water pump and refrigerator/freezer
2. Junction box with connectors for DC to AC interface
3. Kilowatt-hour meter on power consumption
4. Ampere-hour meter to monitor PV production
5. Circuit breaker box with eight circuits
6. Monitoring and switching panel
   DC voltmeter and ammeter
   AC voltmeter and ammeter
   Circuit switches
   Strip chart recorder for measuring production and consumption patterns

PV CONVERSION MATERIALS LIST

26 each 35-watt solar modules
PV array mounting hardware
Solar array regulator
Battery bank
Wiring
Breaker box
Junction box
Power Misers, three each
12-volt DC voltmeter
12-volt DC ammeter
120-volt AC voltmeter
120-volt AC ammeter
Ampere-hour meter
200-watt inverter
500-watt frequency-regulated inverter
500-watt inverters, three each

(Design Note: A more efficient cost-effective approach may be through the use of a Best inverter with the automatic load sensing accessory.)

APPENDIX C
RECHARGEABLE BATTERIES
& The 12-Volt House Battery
By Steve Willey

RECHARGING NI-CAD OR GEL-CELL
RADIO OR FLASHLIGHT BATTERIES

Portable battery-powered devices can be recharged from 12 volts DC if:

1. the voltage of the battery pack is *less than or equal to* 12 volts. [If higher, it will not charge but rather run down into the house battery, since voltage (pressure) flows to lower.]
2. the amount of current flowing from the 12-volt house outlet into the portable batteries can be limited to a very small amount. [Most penlight, C and D size cells can only be recharged at 0.020 to 0.200 amps, usually measured in milliamps (1/1000 amp). So, 20 to 200 milliamps (mA) is usual.]

Determine the voltage of the device:

1. It may be listed on the label.
2. It may be listed on a charger that came with it for AC charging.

3. Open the device and see the battery type: each penlight cell or C or D cell is about 1.5 volts.
4. Measure it with a meter (volts).

Determine the current (rate at which the battery can be recharged without damaging it):

1. It may be on the label or instructions if the device originally came with built-in batteries.
2. The charger that came with it, if any, may list its maximum current. Or, using the charger, cut one wire and insert a meter to measure current, 500 mA scale.
3. Penlights charge at about 89 mA; C or D cells about 125 mA; lantern-type 6-volt gel-cells about 200 mA.

The easiest way to limit the current to your determined value is to make a cord and connect this device to your 12-volt outlet. (This may require a special plug, which you can make by cutting cord from the AC charger that may have come with the unit, as with cordless shavers and electric drills.) Place a flashlight bulb in series with one of the wires in this cord. The bulb will limit the current that can pass. Use a 12-volt bulb so that if the rechargeable battery is completely dead when first connected, the bulb will withstand the difference in voltage between the house (12) and the rechargeable portable (0 if dead). Radio Shack sells bulbs with designated voltage and current.

Select a 12-volt bulb that closely matches the current mA rating (within 50%) as follows:

Battery Current of	Recharging a Battery (current mA Rating)			
	3-volt	6-volt	7.5-volt	9-volt
50 mA	75	100	150	200
100 mA	125	200	300	400
150 mA	175	300	450	600 (0.6 amp) } automotive
200 mA	250	400	600 (0.6 amp)	800 (0.8 amp) } bulbs
250 mA	300	500	750 (0.75 amp)	1 amp

USE AND CARE OF PORTABLE
RECHARGEABLE BATTERIES

Ni-cad batteries are expensive, but can be recharged over 500 times. In the long run, they are inexpensive if properly maintained. They will quit working temporarily if their temperature goes below freezing, so they may not be the wisest choice for flashlights in winter. It is best to run them all the way down before recharging them slowly over a 24-hour period. Their tendency is to form a "memory." If continually recharged or run down partially, they will begin to limit their reserve power to what you have been using between charges. So cycle them fully each time. If you must charge them fully (say, for a trip) and they are not yet empty, it's OK once in a while, but not all the time.

Gel-cells are different—lead-acid batteries with gelled electrolyte instead of liquid. Sears has a 6-volt lantern cell, the kind with two spring contacts on top, that is excellent in a flashlight. This type can be recharged about 250 times. But, like car batteries and unlike the ni-cad, *don't* run them all the way down. Always recharge after use or at least once a month. Keep gel-cells fully charged. They charge best slowly, four hours for each hour's use. They're good. Your light is always ready, always fully bright.

OPERATING PORTABLE RADIOS ON A
12-VOLT HOUSE POWER SYSTEM
(when the unit operated needs less than 12 volts:
6, 9, 7.5, etc.)

Adapters are available to run tape decks and radios from a 12-volt auto battery, and can be set for lesser voltages. Sometimes they are designed for a lesser voltage—usually 6, 7.5 or 9. Do not run these radios directly off 12 volts if they require a lower voltage. You may get away with it for a while with a 9-volt unit, or for a few minutes with a 6-volt unit, but when they fail from overvoltage, repairs will be major.

Adapters usually come with cigarette lighter socket plugs. You can install a socket in the house for this purpose, but it's better to solder a wire onto the voltage converter and be able to plug it into the regular house outlets.

Another possibility is to make the converter. It will cost less and will likely be a better unit, made to suit your specifications.

Do not leave converters plugged in when you turn off the radio; some continue to consume power (you can tell if they are warm).

It may be more worthwhile to put rechargeable (ni-cad or gel-cell) batteries in the radio and operate it from the batteries, recharging them overnight for free by connecting them to your house 12-volt outlet.

OPERATION OF STEREO EQUIPMENT
FROM 12-VOLT DC POWER SYSTEMS

The easiest way to operate a stereo on 12 volts is to use an automotive-type radio and tape player and good-quality speakers, 8-inch diameter minimum, preferably 12-inch. You might as well get the best you can afford because speaker quality is compromised in auto stereos. Some are junk, 4 watts power with 8% distortion admitted. Twelve-inch coaxial speakers (available, unmounted, through Radio Shack) are an economical way to good sound. They must be mounted in a box, cabinet, or closet door to sound as they should. The air movement caused by motion of the cone will simply go around back to fill the void if not enclosed.

Another way to operate a stereo is a little more trouble but results in far better quality. Transistor equipment uses DC. A regular plug-in AC stereo system converts the AC to lower voltages of DC. If the stereo you already have uses only a few reasonably low DC voltages in its circuits, it is possible to supply these to it directly by injecting the right voltage into the circuit where needed. A lot of sets work off one or two basic voltages and from these derive several others. Thus, supplying it with the basic voltages needed runs it the same as AC does. And very small amounts of power are needed so that the voltages over 12 volts DC can be supplied by additional rechargeable batteries (motorcycle batteries or lantern-type gel-cells). These will last for 1 to 4 months and then can be recharged overnight from your deep-cycle main batteries. The only hitch is if the stereo needs many or high voltages, like +80 or −80 volts DC, thus requiring additional batteries. Mine takes +24 and −24.

TV is more of an energy consumer. Only 12-volt sets should be considered, and even they have a tube in them—the picture tube. It has a little heater inside that glows red, and this takes as much power as the rest of the set. A black and white set uses 1 to 2 amps, which is not bad—just about the same as a good 12-volt reading light. Color TV is available for 12 volts but takes about five times the power. TV has sound, picture and color. To add that last aspect, the power needs go up 500%. That means on limited power you can watch black and white TV for 5 hours for every 1 hour you run a color set. Zenith and Sony make the best 12-volt sets.

Turntables that use DC motor belt drive frequently use a 12-volt motor and thus a wiring change allows operation directly from the batteries. Some cassette stereo recording decks are also equipped with 12-volt motors. Wiring modifications need to be done by a person familiar with such procedures, and a circuit diagram is very helpful. I can't recommend specific model numbers to shop for because they keep changing.

Some of the larger portable cassette/am/fm units can be powered from 12 volts, or 9 or 6 volts derived from your 12-volt system through an adapter. CB and amateur radio equipment is available for use directly from 12-volt DC.

APPENDIX D
REFRIGERATION NOTES
For Low-Consumption
Electric Systems
By Steve Willey

Most commercially available refrigerators are designed to be small on the outside and big on the inside. This means less insulation is used and more power is required to keep cool. The front-opening door causes cold air loss at each opening, more motor running to re-cool, and added frost. An energy-efficient refrigerator or freezer should have plenty of insulation on all surfaces, perhaps up to a foot of foam type, and should have a tight-fitting, equally insulated top-opening door. The unit should be located in a cool, shaded area with free airflow to the heat-dispersing parts. The motor and radiator should not be boxed in the same enclosure as the cold storage area, and certainly not underneath where it would heat the bottom.

One type available is the 12-volt compressor refrigerator made for camping and boating. These are more expensive than 120-volt units of the same size ($350 to $700). They are small and light, and so usually are not well insulated; most have a front-opening door. Insulation can be added but since the compressor is often mounted too close below or behind, it is relatively ineffective. Note: we are talking about an electric unit, not a gas-electric combination which usually has an electric heating element (not a gas flame) instead of a compressor and uses so much power that it cannot be considered.

A better option is the compressor unit and icebox mechanics available separately—you build an insulated chest with top opening. Low-cost freight-damaged sheet-metal enamel bathtubs make good inner liners (with drain). Attach wood or plastic rails to the inside and slide wire refrigerator or oven racks for the top shelf, with access to the bottom. Cover the outside with wood panels and elevate the whole thing to a convenient loading height. Cold Machine DCM-12 is a 12-volt compressor that takes over 5 amps running, and runs 50% of the time in a stock icebox—less if used in design described; it will cool 6 to 10 cubic feet (from Alternative Energy Engineering, Redway, California). A better unit, made for built-in refrigerators on yachts, uses cold storage material in the icebox so that it runs only 2 to 4 hours each day, all at once, if used in a super-insulated top-opening chest as described (from Magna Kold, Costa Mesa, California). J. C. Whitney offers a do-it-yourself option for 3.5 and 6 cubic foot units. These can be used to convert ice-cooled cabinets in trailers to refrigerators, and each could do more volume in your own super-insulated box.

Another unit available is the 12-volt thermocouple from Koolatron. A metal junction gets hot on one side and stays cold on the other as current is passed through. Based on the thermoelectric principle, the unit has no moving parts except a small fan to blow the heat away. It is very durable and operates in any position. Unfortunately, it requires 4 amps with 50% running time and cools only about 1 cubic foot. It looks like an ice cooler for camping. Average hourly current consumed would be 2 amps. Also available as a separate unit to use in your own super-insulated box is the cooling panel and fan, good for 5 cubic feet if you use 6 inches of foam all around.

In northern country, you will want to build your box on a porch or an outside unheated area against an inside wall. In winter the unit will not need to run since the temperature inside will be at or below that of its motor thermostat setting. In places where it might get *too* cold, you can run a plastic pipe through the wall into the house and rig an automotive choke as an automatic damper in the refrigerator to control the flow of a little warmth from the house. Some folks use two rocks or jugs of water. One is in the refrigerator, one is warm in the house. Whenever the temperature in the refrigerator gets too low, they just swap rocks or waterjugs for a boost to the refrigerator. With a super-insulated box this works pretty well. I've seen such boxes stay impressively cool in summer with no motor—with just two buckets of cold well water added each morning. Wind-pumped well water can run to the box before going to tanks. Just insulate the pipe before the box and make several passes around the inside. Plastic pipe is best when passing through the insulation. Also use plastic for the drain on the tub or liner, because metal conducts heat out. Some folks have elaborate winter ice storage systems in huge super boxes. In the past, large units were used for entire towns.

Some years back I received a Department of Energy appropriate technology grant to test a refrigeration invention, a direct solar-powered unit, marketed by Zeopower Corporation. The system uses the principle that evaporating water absorbs heat, or becomes cold. Water in a vacuum will evaporate fast enough at room temperature to form ice. In fact, it will appear to boil. Water boils at lower temperatures at higher altitudes; that is, as pressure is reduced. In a complete vacuum the water boils away at room temperature and, in the process, absorbs so much heat (creates so much cold) that, rather than evaporate, much of the remaining water becomes solid ice. This process would stop itself quickly if sealed in a container under vacuum, because that space above the water surface would soon become saturated with water vapor and the vapor pressure would act like air pressure on the water, effectively eliminating the vacuum.

Zeopower connected this water container by pipe to Zeolite, a silica gel substance that absorbs water vapor with great enthusiasm and can be dried by baking in the oven. (Zeolite is used in thermopane windows to prevent inside fogging.) The Zeopower refrigerator, then, is a jug of water in an insulated box. It is connected to a solar panel containing Zeolite—completely evacuated of air pressure inside the jug, pipe and panel. The water boils and freezes in that vacuum and the vapor is absorbed by the Zeolite in the panel, reducing the vapor pressure against the water. The process continues until the remaining water is frozen. During the day the sun heats the black and glass solar panel and forces the water out of the Zeolite. The water condenses in the pipe, which includes a small radiator to cool the water to room temperature, and runs back into the insulated glass jug and onto the block of ice. At night the same water again partially evaporates, partially freezes, and after three such day/night cycles, nearly all the water in the jug is frozen (enough, it is claimed, to hold over three days with no sun).

Quite an idea, considering that it is powered by sunshine, that there are no mechanical parts and, if well made, it should last a very long time. Now, for the negative aspects. You may have already realized that this unit freezes its own ice and from that ice works like any icebox. It won't keep ice cream or make ice for drinks. Zeopower is investigating another liquid that will make ice in the jug not at 32°, but at 24° so that ice-making may be possible.

My unit arrived with concealed freight damage wherein the somewhat underengineered frame failed in transit and the water jug was broken, releasing the vacuum. After several years working with the manufacturer, DOE and freight companies, it is finally repaired and is being tested. The icebox is very poorly insulated, the door seal is also poor, and the gallon jug takes up most of the space inside. This is a test model and, at a cost of $1900, will not be of interest in its present form. While Zeopower claims the price will be $500 someday, they are now considering using this method to air-condition large buildings and so are not working on the refrigerator at present.

A promising development in low-power refrigeration was recently reported in *The PV Network News*. Sun Frost, a small firm in northern California, has begun custom-building 17-cu ft vertical refrigerators which consume 180 kWh per year or 41 ampere-hours per day. The most efficient unit at present of comparable size consumes five times more energy. For details and prices contact Larry Schlussler at Sun Frost (see Appendix G).

APPENDIX E
CURRENT CARRYING CAPACITY OF COPPER WIRE

The ratings in the following tabulations are those permitted by the National Electrical Code for flexible cords and for interior wiring of houses, hotels, office buildings, industrial plants, and other buildings.

The values are for copper wire. For aluminum wire the allowable carrying capacities shall be taken as 84% of those given in the table for the respective sizes of copper wire with the same kinds of covering.

Size A.W.G.	Area Circular (mils)	Diameter of Solid Wires (mils)	Rubber Insulation (amps)	Varnished Cambric Insulation (amps)	Other Insulations and Bare Conductors (amps)
24	404	20.1	−	−	1.5
22	642	25.3	−	−	2.5
20	1,022	32.0	−	−	4
18	1,624	40.3	3*	−	6**
16	2,583	50.8	6*	−	10**
14	4,107	64.1	15	18	20
12	6,530	80.8	20	25	30
10	10,380	101.9	25	30	35
8	16,510	128.5	35	40	50
6	26,250	162.0	50	60	70
5	33,100	181.9	55	65	80
4	41,740	204.3	70	85	90
3	52,630	229.4	80	95	100
2	66,370	257.6	90	110	125

Note: 1 mil = 0.001 inch.

*The allowable carrying capacities of No. 18 and 16 are 5 and 7 amperes, respectively, when in flexible cord.

**The allowable carrying capacities of No. 18 and 16 are 10 and 15 amperes, respectively, when in cords for portable heaters. Types AFS, AFSI, HC, HPD and HSJ.

CONVERSION FACTORS

To Change	Into	Multiply by
BTU	cal	252
BTU	joules	1,055
BTU	kcal	0.252
BTU	kWh	2.93×10^{-4}
BTU ft^{-2}	langleys (cal cm^{-2})	0.271
cal	BTU	3.97×10^{-5}
cal	ft-lb	3.09
cal	joules	4.184
cal	kcal	0.001
cal min^{-1}	watts	0.0698
cm	inches	0.394
cc or cm^3	in.^{3}	0.0610
ft^3	liters	28.3
in.^{3}	cc or cm^3	16.4
ft	m	0.305
ft-lb	cal	0.324
ft-lb	joules	1.36

ft-lb	kg-m	0.138
ft-lb	kWh	3.77×10^{-7}
gal	liters	3.79
hp	kW	0.745
inches	cm	2.54
joules	BTU	9.48×10^{-4}
joules	cal	0.239
joules	ft-lb	0.738
kcal	BTU	3.97
kcal	cal	1,000
kcal min^{-1}	kW	0.0698
kg-m	ft-lb	7.23
kg	lb	2.20
kW	hp	1.34
kWh	BTU	3,413
kWh	ft-lb	2.66×10^6
kW	kcal min^{-1}	14.3
langleys (cal cm^2)	BTU ft^{-2}	3.69
langleys min^{-1} (cal cm^{-2} min^{-1})	watts cm^{-2}	0.0698
liters	gal	0.264
liters	qt	1.06
m	ft	3.28
lb	kg	0.454
qt	liters	0.946
cm^2	ft^2	0.00108
cm^2	in.^{2}	0.155
ft^2	m^2	0.0929
m^2	ft^2	10.8
watts cm^{-2}	langleys min^{-1} (cal cm^2)	14.3

APPENDIX G
MANUFACTURERS & SUPPLIERS

The most practical voltage to operate a PV system is 12-volt DC, and most manufacturers of PV panels produce 12-volt DC modules. Both standard automotive and recreational vehicle equipment operate on 12-volt DC, making most equipment you'll need universally available. When gathering materials, be sure to check local suppliers—hardware stores, battery dealers, etc., and, of course, salvage yards. For a more complete listing of manufacturers and suppliers, see *Solar Census—Photovoltaics Edition.*

INVERTER MANUFACTURERS AND SUPPLIERS

Best Energy Systems
P.O, Box 280
Necadah, WI 54646
(608) 565 7200

Dynamote Corp.
1200 W. Nickerson
Seattle, WA 98119
(206) 282 1000

Dytek Laboratories, Inc.
Airport International Plaza
Bohemia, NY 11716
(516) 567 8500

Heart Interface
1626 South 341st Place
Federal Way, WA 98003
(800) 732 3201

Honeywell Motor Products
P.O. Box 106
Rockford, IL 61105
(815) 966 3600

Topaz Electronics
3855 Riffin Road
San Diego, CA 92123
(714) 565 8363

Tripp-Lite Manufacturing Co.
500 N. Orleans
Chicago, IL 60610
(312) 329 1777

Wilmore Electronics Co., Inc.
P.O. Box 2973
West Durham Station
Durham, NC 27705
(919) 489 3318

Also see—Alternative Energy Engineering; The Earth Store; Radio Shack; U.S. General; J. C. Whitney

Grid-Connect Inverters

American Power Conversion
89 Cambridge Street
Burlington, MA 01803
(617) 273 1570

Helionetics, Inc.
DECC Division
17312 Eastman Street
Irvine, CA 92714
(714) 546 4731

Windworks, Inc.
Route 3, Box 44A
Mukwonago, WI 53149
(414) 363 4088

PHOTOVOLTAICS MANUFACTURERS

Acurex
485 Clyde Avenue
Mountain View, CA 94042
(415) 964 3200

Free Energy Systems, Inc.
P.O. Box 3030
Lenni, PA 19052
(215) 583 4780

Ametek, Inc.
Station Square Two
Paoli, PA 19301
(215) 647 2121

Mobil Solar Energy Corporation
16 Hickory Drive
Waltham, MA 02254
(617) 890 1180

Applied Solar Energy Corporation
15751 East Don Julian Road
P.O. Box 1212
City of Industry, CA 91746
(213) 968 6581

Photon Energy
13 Founders Boulevard
El Paso, TX 79906
(915) 779 7774

ARCO Solar, Inc.
21011 Warner Center Lane
Box 4400
Woodland Hills, CA 91365
(213) 700 7000

Silicon Sensors, Inc.
Highway 18 East
Dodgeville, WI 53533
(608) 935 2707

Silonex Inc.
331 Cornelia Street
Plattsburgh, NY 12901
(518) 561 3160

Chronar Corporation
P.O. Box 177
Princeton, NJ 08542
(609) 587 8000

Solar Power Corporation
20 Cabot Road
Woburn, MA 01801
(617) 935 4600

Energy Conversion Devices, Inc.
1675 W. Maple Road
Troy, MI 48084
(313) 280 1900

Solarex Corporation
1335 Piccard Drive
Rockville, MD 20850
(202) 948 0202

Solavolt International, Inc.
3646 E. Atlanta
P.O. Box 2934
Phoenix, AZ 85062
(602) 231 6408

Solec International, Inc.
12533 Chadron Avenue
Hawthorne, CA 90250
(213) 970 0065

Solenergy Corporation
171 Merrimac Street
Woburn, MA 01801
(617) 938 0563

Sol/Los, Inc.
1519 Comstock Avenue
Los Angeles, CA 90024
(213) 203 0728

SunWatt Corporation
Route 2
English, IN 47118
(812) 338 3163

Tideland Signal Corp.
4310 Director's Row
P.O. Box 52430
Houston, TX 77052
(713) 681 6101

United Energy Corp.
420 Lincoln Centre Drive
Foster City, CA 94404
(415) 570 5011

SUPPLIERS

Airborne Sales
P.O. Box 2727
Culver City, CA 90230
(catalog, surplus equipment,
including 12-volt DC)

Alder/Barbour Marine Systems, Inc.
43 Lawton Street
New Rochelle, NY 10801
(manufactures 12-volt DC refrigeration systems)

Alternative Energy Engineering
P.O. Box 339
Briceland Star Route
Redway, CA 95489
(catalog, PV equipment, inverters,
refrigerators & other appliances,
wiring needs, etc.)

Appropriate Technology
    See Dave Lukenback

Backwoods Cabin Electric Systems
    See Steve Willey

Val Bertoia
644 Main Street
Bally, PA 19503
(manufactures high-quality wind-
chargers)

Bliss Marine
Route 128 at Exit 61
Dedham, MA 02026
(marine equipment catalog,
some 12-volt DC)

Cole Hersee
20 Old Colony Avenue
Boston, MA 02127
(automotive DC equipment)

Windy Dankoff
Windlight Workshop
P.O. Box 6015
Santa Fe, NM 87502
(supplier of DC motors and
other 12-volt hardware)

The Earth Store
P.O. Box 679
North San Juan, CA 95960
($2 catalog of PV and home
appliances, inverters, etc.)

Kenneth Foster
1742 Dowd Street
St. Louis, MO 63136
(solar cells)

Harbor Freight Salvage Co.
3491 Mission Oaks Boulevard
Camarillo, CA 93010
(12-volt DC lights)

Hi-Craft Metal Products
606 W. 184th Street
Gardena, CA 90247
(fluorescent lights)

Independent Power Company
12340 Tyler Foote Road
Nevada City, CA 95959
(catalog of PV equipment, batteries,
appliances, etc.)

IOTA Engineering
4700 S. Park Avenue, Suite 8
Tucson, AZ 85714
(manufactures high-quality
fluorescent lights and ballasts)

Greg Johanson
Solar Electrical Systems
9742 Cactus Avenue
Chatsworth, CA 91311
(designs, distributes &
installs complete PV systems)

Noel Kirkby
Solar Electric Systems
Fleming Springs
Box 1562
Cave Creek, AZ 85331
(supplier of PV equipment
for the RV'er)

Koolatron
56 Harvester Avenue
Batavia, NY 14020
(refrigerators)

Lafayette Electronics
(electronics supplies: meters,
fuses, low-power equipment)

William Lamb Co.
10615 Chandler Blvd.
North Hollywood, CA 91601
(general PV supplier)

Jim Lavandier
Box 4869
San Francisco, CA 94101
(12-volt DC record players,
amplifiers, etc.)

The Light Fantastic
P.O. Box 564
Garberville, CA 95440
(quartz bulbs and fixtures)

Dave Lukenback
Appropriate Technology
Route 1, Box 269
Ava, MO 65608
(manufactures solar trackers)

Magna Kold
1760 Morovia Avenue
Costa Mesa, CA 92627
(refrigerators)

March Pumps
1819 Pickwick Avenue
Glenview, IL 60025
(manufactures PV pumps)

McLean Electronics Inc.
10810 Talbert Avenue
Fountain Valley, CA 92708
(fluorescent lights)

Mechanical Products, Inc.
P.O. Box 729
Jackson, MI 49204
(12-volt DC circuit breakers)

National Trailer Stores
Harry Gramig
610 W. Florence Avenue
Englewood, CO 90301
(Dometic gas refrigerators
from Sweden)

Norcold Inc.
1510 Michigan Street
Sidney, OH 45365
(manufactures DC and RV
refrigerators)

Now Devices
Unit E
7975 E. Harvard Avenue
Denver, CO 80231
(manufactures PV regulators)

Bill Perleberg
24110 US Hwy 40
Golden, CO 80401
(12-volt DC water heater
elements)

Princess Auto
Box 1005
475 Panet Road
Winnipeg, Manitoba
Canada R3C 2W7
(catalog of tools, electrical and
other equipment for the home-
steader and farmer)

Radio Shack
(electronics supplies: meters, fuses,
connectors, low-power equipment)

REC, Inc.
530 Constitution Ave
Camarillo, CA 93010
(fluorescent lights)

Earl Schmidt
5 Marion Avenue
Albany, NY 12203
(designs and installs hydroelectric power plants)

Sears
(catalog of mobile home, RV and
camping supplies)

Solar Electric Systems
 See Noel Kirkby

Solar Electrical Systems
 See Greg Johanson

Solar Electronics
156 Drakes Lane
Summertown, TN 38483
(PV supplier/installer)

Solar Marine Systems, Inc.
1095 Normington Way
San Jose, CA 95136
(PV boat supplies)

Solar Usage Now
Box 306
Bascom, OH 44809
($5 catalog, all types of solar
and energy-saving equipment)

Solarwest Electric
232 Anacapa Street
Santa Barbara, CA 93101
($3 catalog of ARCO Solar PV
equipment and systems)

Solar Works!
 See Paul Wilkins

Specialty Concepts
9025 Eton Avenue, Suite D
Canoga Park, CA 91304
(PV regulators, meters and
junction boxes)

Standard Solar Collectors, Inc.
1465 Gates Avenue
Brooklyn, NY 11227
(PV refrigerators)

Sun Frost
Larry Schlusser
725 Bayside
Arcata, CA 95521
(refrigerators)

Surplus Center
1000-1015 West "O" Street
P.O. Box 82209
Lincoln, NE 68501
(surplus equipment catalog,
wide range of hardware)

Tensen Co., Inc.
304 S.E. Second
Portland, OR 97214
(manufacturer of 12-volt DC
chainsaws)

U.S. General
100 Commercial Street
Plainview, NY 11803
(tool catalog with inverters, etc.)

Vanderbeck Lightning Rod Company
P.O. Box 405
Ho-ho-kus, NJ 07423
(lightning protection equipment
for solar arrays)

Western Solar Refrigeration
715 J Street
San Diego, CA 92101
(PV refrigerators)

J. C. Whitney
1917-19 Archer Avenue
P.O. Box 8410
Chicago, IL 60680
(automotive equipment catalog;
12-volt DC, lights, pumps, meters,
cables, appliances)

Paul Wilkins
Solar Works!
Route 2, Box 274
Santa Fe, NM 87501
(manufactures and distributes
Charge-a-Stat regulator and
control boards for PV systems)

Steve Willey
Backwoods Cabin Electric
  Systems
8530 Rapid Lightning Creek Road
Sandpoint, ID 83864
(manufactures and distributes
control and meter boards for
PV systems)

Windlight Workshop
  See Windy Dankoff

Zomeworks
P.O. Box 25805
Albuquerque, NM 87125
(manufactures freon-driven solar
trackers for PV systems)

INFORMATION SOURCES

BOOKS

Better Use of . . . by Michael Hackleman, $9.95, Earthmind, 4844 Hirsch Road, Mariposa, CA 95338. (This excellent book is a must. Michael's books on alternative energy, electric vehicles, and wind power also have sections useful to the PV do-it-yourselfer.)

Basic DC Circuits by Franklin Swan and Warren Palmer, $1.95, Radio Shack, Ft. Worth, TX 76107. (Basic book on DC theory and circuitry.)

How To Be Your Own Home Electrician by George Daniels, $1.95, Popular Science Publishing Co., Inc., 355 Lexington Avenue, New York, NY 10017. (How-to book on house wiring.)

How To Be Your Own Power Company by Jim Cullen, $10.95, Van Nostrand Reinhold Co., 135 W. 50th Street, New York, NY 10020. (Examines low voltage, direct current, power generating systems.)

How To Design An Independent Power System, $6, Best Energy Systems, Route 1, Box 280, Necedah, WI 54646. (Manufacturer explains the use of large inverters.)

Photovoltaics by Paul Maycock and Edward Stirewalt, $9.95, Brick House Publishing Co., Inc., 34 Essex Street, Andover, MA 01810. (Overviews the PV industry and its development.)

Practical Photovoltaics by Richard Komp, $16.95, **aatec publications**, P.O. Box 7119, Ann Arbor, MI 48107. (Updated edition; module and array construction and installation; theory, politics, history and future of PV. Companion to *The Solar Electric Home*.)

The Solar Boat Book by Pat Rand Rose, $10.95, Aqua-Sol Enterprises, P.O. Box 18646, Ft. Worth, TX 76118. (The only book on solar energy use for boaters.)

Solar Census — Photovoltaics Edition, $14.95, **aatec publications**, P.O. Box 7119, Ann Arbor, MI 48107. (Directory of over 500 manufacturers, suppliers, designers, educators, researchers.)

The Solarex Guide to Solar Electricity by The Staff of Solarex, $6.95, Solarex Corporation, 1335 Piccard Drive, Rockville, MD 20850. (Good beginning book on photovoltaics.)

Wind/Solar Energy by Edward M. Noll, $12.95, Howard W. Sams & Co., Box 7092, Indianapolis, IN 46206. (Complete info on wind & PV for radiocommunications and low-power electric systems.)

BOOKLETS

Batteries

"Battery Service Manual," $2, The Battery Council International, 111 East Wacker Drive, Chicago, IL 60601. (The best source of battery service information—a must.)

C&D Batteries, 3043 Walton Road, Plymouth Meeting, PA 19462. (Information booklet on storage batteries.)

"Facts About Storage Batteries," ESB Brands, Inc., P.O. Box 6949, Cleveland, OH 44101.

"The Storage Battery," Exide, 101 Gibraltar Road, Horsham, PA 19044. (Examines lead-acid batteries.)

"Stationary Battery Installation and Operating Instructions," Gould, Inc., Industrial Battery Division, 2050 Cabot Boulevard West, Langhorne, PA 19047.

"Capturing the Sun—Batteries and Solar Energy Storage," Lead Industries Association, 292 Madison Avenue, New York, NY 10017.

"Golf Cart Battery Maintenance Manual," SGL Industries, Inc., 14650 Dequindre, Detroit, MI 48212.

Other

"Energy from Solar Cells," State of Arkansas Energy Office, 960 Plaza West, Little Rock, AR 72205. (Describes the large-scale solar cell project to power Mississippi County Community College, Blytheville, Arkansas.)

"Site Selection for Solar Suitability" by Joel Davidson, The Office of Human Concern, P.O. Box 756, Rogers, AR 72756. (General information for siting any solar device.)

PERIODICALS

AERO Sun Times, Alternative Energy Resource Organization, 424 Stapleton Building, Billings, MT 59101.

Alternate Energy Transportation Newsletter, Ed Campbell, c/o Electric Vehicle Consultants, 327 Central Park West, New York, NY 10025. (Information on electric cars.)

Alternative Sources of Energy, 107 South Central, Milaca, MN 56353. (Excellent coverage of all phases of alternative energy.)

ARCO Solar News, Box 4400, Woodland Hills, CA 91365. (Free.)

The Mother Earth News, P.O. Box 70, Hendersonville, NC 28739. (Occasional articles on solar electricity.)

The PV Network News, c/o Joel Davidson, 10615 Chandler Boulevard, North Hollywood, CA 91601. ($10 per year for newsletter, basic information packet, and membership in the PV Network.)

PV News, PV Energy Systems, 2401 Childs Lane, Alexandria, VA 22308. (Paul Maycock's monthly newsletter.)

Renewable Energy News, P.O. Box 32226, Washington, DC 20007. (Monthly newspaper examines alternative energies worldwide.)

Solar Age Magazine, Church Hill, Harrisville, NH 03450. (The most important magazine in the solar field.)

Solar Energy Digest, c/o William Edmondson, P.O. Box 88, Ocotilla, CA 92259. (Covers PV, solar thermal, bio and other energies.)

Solar Utilization News, P.O. Box 3100, Estes Park, CO 80517. (General solar newspaper with some articles on PV.)

Solarex Newsletter, 1335 Piccard Drive, Rockville, MD 20850. (Free.)

Wind Power Digest, P.O. Box 306, Bascom, OH 44809.

RESOURCE PEOPLE

American Solar Energy Society, Inc., 1230 Grandview Avenue, Boulder, CO 80302. (The largest public and professional solar group.)

Stephen Cook, Director, North Arkansas Community College Energy Center, Pioneer Ridge, Harrison, AR 72601. (Good source of wind/PV combination systems information; president of CompuSOLAR, a solar computer software and educational firm.)

David Copperfield, Well-Being Productions, Oak Valley Star Route, Camptonville, CA 95922. (David has written several practical how-to booklets for the 12-volt DC user on converting washers, turntables, dual battery systems, etc.)

Windy Dankoff, Windlight Workshop, P.O. Box 6015, Santa Fe, NM 87501. ($2 and SASE for information on PV water pumping systems; catalog.)

Joel Davidson, 10615 Chandler Boulevard, North Hollywood, CA 91601, (213) 980 6248. ($10 for membership in the PV Network and subscription to *The PV Network News;* consultant; assists in locating hard-to-find PV components.)

Greg Johanson, Solar Electrical Systems, 9640 Wilbur Avenue, Northridge, CA 91324, (213) 993 7597. ($5 booklet on systems and controls.)

Noel Kirkby, Solar Electric Systems, Fleming Springs, Box 1562, Cave Creek, AZ 85331. ($5 for Noel's information packet geared toward the recreational vehicle PV user.)

Richard Komp, Skyheat, Route 2, English, IN 47118, (812) 338 3163. (Lectures and conducts hands-on PV workshops both at Skyheat and nationwide; SASE for information.)

Ted Landers, Perennial Energy, Inc., P.O. Box 15, Dora, MO 65637. (Information on induction motors for grid-connect.)

Paul Wilkins, Solar Works!, Route 2, Box 274, Santa Fe, NM 87501. ($5 for Paul's information packet.)

Steve Willey, Backwoods Cabin Electric Systems, Route 1, Box 461X, Sandpoint, ID 83864. ($5 for Steve's packet of wind and PV system information.)

APPENDIX I
NATIONAL ELECTRICAL CODE: ARTICLE 690
Solar Photovoltaic Systems

The following is the proposed National Electrical Code for photovoltaic-powered buildings. The requirements are quite simple and make a lot of sense, so even if you are not required to build to code, it is wise to do so to ensure that your installation is safe and efficient. Remember, the codes given here are preliminary and may differ slightly in their final approved form.

A. GENERAL

690-3 Other Articles. Wherever the requirements of other Articles of this Code and Article 690 differ, the requirements of Article 690 shall apply.

690-4 Installation.

(a) Photovoltaic System. A solar photovoltaic system shall be permitted to supply a building or other structure in addition to any service(s) of another electricity supply system(s).

(b) Conductors of Different Systems. Photovoltaic source circuits and photovoltaic output circuits shall not be contained in the same raceway, cable tray, cable, outlet box, junction box or similar fitting as feeders or branch circuits of other systems.

Exception: Where the conductors of the different systems are separated by a partition or are connected together.

(c) Module Connection Arrangement. The connection to a module or panel shall be so arranged that removal of a module or panel from a photovoltaic source circuit does not interrupt a grounded conductor to another photovoltaic source circuit.

B. CIRCUIT REQUIREMENTS

690-7 Maximum Voltage.

(a) Voltage Rating. In a photovoltaic power source and its direct current circuits, the voltage considered shall be the rated open-circuit voltage.

(b) Direct Current Utilization Circuits. The voltage of direct current utilization circuits shall conform with Section 210-6.

(c) Photovoltaic Source and Output Circuits. Photovoltaic source circuits and photovoltaic output circuits which do not include lampholders, fixtures or standard receptacles shall be permitted up to 600 volts.

(d) Circuits Over 150 Volts to Ground. In one- and two-family dwellings, live parts in photovoltaic source circuits and photovoltaic output circuits over 150 volts to ground shall not be accessible while energized, to other than qualified persons.

690-8 Circuit Sizing and Current.

(a) Ampacity and Overcurrent Devices. The ampacity of the conductors and the rating or setting of overcurrent devices in a circuit of a solar photovoltaic system shall not be less than 125 percent of the current computed in accordance with (b) below. The rating or setting of overcurrent devices shall be permitted in accordance with Section 240-3, Exception No. 1.

Exception: Circuits containing an assembly together with its overcurrent device(s) that is listed for continuous operation at 100 percent of its rating.

(b) Computation of Circuit Current. The current for the individual type of circuit shall be computed as follows:

(1) Photovoltaic Source Circuits. The sum of parallel module current ratings.

(2) Photovoltaic Output Circuit. The photovoltaic power source current rating.

(3) Power Conditioning Unit Output Circuit. The power conditioning unit output current rating.

Exception: The current rating of a circuit without an overcurrent device, as permitted by the Exception to Section 690-9(a), shall be the short-circuit current, and it shall not exceed the ampacity of the circuit conductors.

690-9 Overcurrent Protection.

(a) Circuits and Equipment. Photovoltaic source circuit, photovoltaic output circuit, power conditioning unit output circuit, and storage battery circuit conductors and equipment shall be protected in accordance with the requirements of Article 240. Circuits connected to more than one electrical source shall have a sufficient number of overcurrent devices so located as to provide overcurrent protection from all sources.

Exception: A conductor in a photovoltaic source circuit, photovoltaic output circuit, or power conditioning unit output circuit having an ampacity not less than the maximum available current under short-circuit or ground-fault conditions with the condition of a shorted blocking diode shall be permitted without an overcurrent device.

FPN: Possible backfeed of current from any source of supply, including a supply through a power conditioning unit into the photovoltaic output circuit and photovoltaic source circuits must be considered in determining whether adequate overcurrent protection from all sources is provided for conductors and modules.

(b) Power Transformers. Overcurrent protection for a transformer with a source(s) on each side shall be provided in accordance with Section 450-3 by considering first one side of the transformer, then the other side of the transformer as the primary.

(c) Photovoltaic Source Circuits. Branch-circuit or supplementary type overcurrent devices shall be permitted to provide overcurrent protection in photovoltaic source circuits. The overcurrent devices shall be accessible, but shall not be required to be readily accessible.

C. DISCONNECTING MEANS

690-13 All Conductors. Means shall be provided to disconnect all current-carrying conductors of a photovoltaic power source from all other conductors in a building or other structure.

690-14 Additional Provisions. The provisions of Article 230, Part H shall apply to the photovoltaic power source disconnecting means.

Exception No. 1: The disconnecting means shall not be required to be suitable as service equipment and shall be rated in accordance with Section 690-17.

Exception No. 2: Equipment such as photovoltaic source circuit isolating switches, overcurrent devices, and blocking diodes shall be permitted ahead of the photovoltaic power source disconnecting means.

690-15 Disconnection of Photovoltaic Equipment. A means shall be provided to disconnect equipment, such as a power conditioning unit, filter assembly and the like from all ungrounded conductors of all sources. If the equipment is energized (live) from more than one source, the disconnecting means shall be grouped and indentified

690-16 Fuses. A disconnecting means shall be provided to disconnect a fuse from all sources of supply if the fuse is energized from both directions and is accessible to other than qualified persons. Such a fuse in a photovoltaic source circuit shall be capable of being disconnected independently of fuses in other photovoltaic source circuits.

690-17 Switch or Circuit Breaker. The disconnecting means for ungrounded conductors shall consist of a manually operable switch(es) or circuit breaker (1) located where readily accessible, (2) externally operable without exposing the operator to contact with live parts, (3) plainly indicating whether in the open or closed position, and (4) having ratings not less than the load to be carried. Where disconnect equipment may be energized from both sides, the disconnect equipment shall be provided with a marking to indicate that all contacts of the disconnect equipment may be live.

690-18 Disablement of an Array. Means shall be provided to disable an array or portions of an array.

FPN: Photovoltaic modules are energized while exposed to light. Installation, replacement, or servicing of array components while a module(s) is irradiated may expose persons to electric shock.

D. WIRING METHODS

690-31 Methods Permitted.

(a) Wiring Systems. All raceway and cable wiring methods included in this Code and such other wiring systems specifically intended and approved for use on photovoltaic arrays shall be permitted with approved fittings and with fittings approved specifically for photovoltaic arrays. Where wiring devices with integral enclosures are used, sufficient length of cable shall be provided to facilitate replacement.

(b) Single Conductor Cable. Type UF single conductor cable shall be permitted in photovoltaic source circuits where installed in the same manner as a Type UF multiconductor cable in accordance with Article 339. Where exposed to direct rays of the sun, cable identified as sunlight-resistant shall be used.

690-32 Component Interconnections. Fittings and connectors which are intended to be concealed at the time of on-site assembly, when listed for such use, shall be permitted for on-site interconnection of modules or other array components. Such fittings and connectors shall be equal to the wiring method employed in insulation, temperature rise and fault-current withstand, and shall be capable of resisting the effects of the environment in which they are used.

690-33 Connectors. The connectors permitted by Section 690-32 shall comply with (a) through (e) below.

(a) The connectors shall be polarized and shall have a configuration that is noninterchangeable with receptacles in other electricity systems on the premises.

(b) The connectors shall be constructed and installed so as to guard against inadvertent contact with live parts by persons.

(c) The connectors shall be of the latching or locking type.

(d) The grounding member shall be the first to make and the last to break contact with the mating connector.

(e) The connectors shall be capable of interrupting the circuit current without hazard to the operator.

690-34 Access to Boxes. Junction, pull and outlet boxes located behind modules or panels shall be installed so that the wiring contained in them can be rendered accessible directly or by displacement of a module(s) or panel(s) secured by removable fasteners and connected by a flexible wiring system.

E. GROUNDING

690-41 System Grounding. For a photovoltaic power source, one conductor of a two-wire system and a neutral conductor of a three-wire system shall be solidly grounded.

Exception: Other methods which accomplish equivalent system protection and which utilize equipment listed and identified for the use shall be permitted.

690-42 Point of System Grounding Connection. The direct current circuit grounding connection shall be made at any single point on the photovoltaic output circuit.

FPN: Locating the grounding connection point as close as practicable to the photovoltaic source will better protect the system from voltage surges due to lightning.

690-43 Size of Equipment Grounding Conductor. The equipment grounding conductor shall be no smaller than the required size of the circuit conductors in systems (1) where the available photovoltaic power source short-circuit current is less than twice the current rating of the overcurrent device, or (2) where overcurrent devices are not employed as permitted in the Exception to Section 690-9(a). In other systems, the equipment grounding conductor shall be sized in accordance with Section 250-95.

690-44 Common Grounding Electrode. Exposed noncurrent-carrying metal parts of equipment and conductor enclosures of a photovoltaic system shall be grounded to the grounding electrode that is used to ground the direct current system. Two or more electrodes that are effectively bonded together shall be considered as a single electrode in this sense.

F. MARKING

690-51 Modules. Modules shall be marked with identification of terminals or leads as to polarity, maximum overcurrent device rating for module protection and with rated (1) open-circuit voltage, (2) operating voltage, (3) maximum permissible system voltage, (4) operating current, (5) short-circuit current and (6) maximum power.

690-52 Photovoltaic Power Source. A marking, specifying the photovoltaic power source rated (1) operating current, (2) operating voltage, (3) open-circuit voltage and (4) short-circuit current, shall be provided at an accessible location at the disconnecting means for the photovoltaic power source.

FPN: Reflecting systems used for irradiance enhancement may result in increased levels of output current and power.

G. CONNECTION TO OTHER SOURCES

690-61 Loss of Utility Voltage. The power output from a utility interactive power conditioning unit shall be automatically disconnected from all ungrounded conductors of the utility system upon loss of voltage in the utility system and shall not reconnect until the utility voltage is restored.

690-62 Ampacity of Neutral Conductor. If a single-phase, 2-wire power conditioning unit output is connected to the neutral and one ungrounded conductor (only) of a 3-wire system or of a 3-phase, 4-wire wye-connected system, the maximum load connected between the neutral and any one ungrounded conductor plus the power conditioning unit output rating shall not exceed the ampacity of the neutral conductor.

690-63 Unbalanced Interconnections.

(a) Single-Phase. The output of a single-phase power conditioning unit shall not be connected to a 3-phase, 3- or 4-wire delta-connected system.

(b) Three-Phase. A 3-phase power conditioning unit shall be automatically disconnected from all ungrounded conductors of the interconnected system when one of the phases opens in either source.

Exception for (a) and (b): Where the interconnected system is designed so that significant unbalanced voltages will not result.

690-64 Point of Connection. The output of the power conditioning unit shall be connected to the supply side of the service disconnect as permitted in Section 230-82, Exception 6.

GLOSSARY

AC—Alternating current; the electric current which reverses its direction of flow. 60 cycles per second is the standard current used by utilities in the U.S.

acceptance angle—The total range of sun positions from which sunlight can be collected by a system.

AH—*see* **ampere-hours**

Air Mass 1 (AM1)—The amount of sunlight falling on the earth at sea level when the sun is shining straight down through a dry clean atmosphere. (A close approximation is the Sahara Desert at high noon.) The sunlight intensity is very close to 1 kilowatt per square meter (1 kW/m^2).

alternating current—*see* **AC**

ampere—A unit of electrical current or the rate of flow of electrons. One volt across one ohm of resistance causes a current flow of one ampere. One ampere equals 6.25×10^{18} electrons per second passing a given point in a circuit; amp.

ampere-hour—A current of one ampere running for one hour.

array—A set of modules or panels assembled for a specific application; may consist of modules in series for increased voltage, or in parallel for increased current, or a combination of both.

battery, marine—A deep-discharge battery used on boats; capable of discharging small amounts of electricity over long time periods.

battery, stationary—For use in emergency stand-by power systems, a battery with long life but poor deep-discharge capabilities.

battery, storage—A secondary battery; rechargeable electric storage unit that operates on the principle of changing electrical energy into chemical energy by means of a reversible chemical reaction. The lead-acid automobile battery is the most familar type.

battery capacity—Expressed in ampere-hours, the total amount of electricity that can be drawn from a fully charged battery until it is discharged to a specific voltage.

battery capacity, available—The total ampere-hours that can be drawn from a battery under specific operating conditions of discharge rate, temperature, initial state of charge, age and cutoff voltage.

battery capacity, energy—The total watt-hours (kilowatt-hours) that can be drawn from a fully charged battery. This varies with temperature, rate, age and cutoff voltage.

battery capacity, installed—The total ampere-hours that can be drawn from a new battery when discharged to the specified maximum depth of discharge.

battery capacity, rated—The manufacturer's conservative estimate of ampere-hours that can be drawn from a new battery under specific conditions.

battery cell—The simplest operating unit in a storage battery; one or more positive electrodes, an electrolyte that permits ionic conduction, one or more negative electrodes, and separators enclosed in a single container.

battery cycle life—The number of cycles to a specified depth of discharge a battery can undergo before efficiency is affected.

battery life—The period when a battery is operating above specific efficiency levels. Measured in either cycles or years, depending on intended use.

blocking diode—A device that prevents current from running backward through an array, thereby draining the storage battery.

BTU—British Thermal Unit; the unit of heat energy sufficient to raise the temperature of one pound of water 1°F.

concentration ratio—The ratio between the area of clear aperture (opening through which sunlight enters) and the area of the illuminated cell.

DC—Direct current; electric current that always flows in the same direction—positive to negative. Photovoltaic cells and batteries are all DC devices.

deep-discharge cycles—Cycles in which a battery is nearly completely discharged.

depth of discharge—The number of ampere-hours withdrawn from a fully charged battery, stated as a percentage of rated capacity.

direct current—*see* **DC**

discharge rate—The current removed from a battery.

electric current—The rate at which electricity flows through an electrical conductor; expressed in amps.

full sun—*see* **Air Mass 1**

grid—The utility network of transmission lines used to distribute electricity.

hybrid system—A system that produces both usable heat (e.g., to heat water) and electricity.

insolation—The amount of sunlight striking a given area.

inverter—A device that converts DC to AC.

inverter, synchronous—A device that converts DC to AC in synchronization with the power line. Excess power is fed back into the utility grid.

kilowatt-hour—Unit of energy used to perform work; kWh.

module—A series string of 32 to 36 cells, producing an open-circuit voltage in bright sunlight of about 18 volts, or 16 volts when producing maximum power. Total current output of a series string is the same as a single cell.

open-circuit voltage—The voltage produced by a solar cell when exposed to standard sunlight conditions, and with no load.

panel—*see* **module**

parallel—In module construction, to increase current output, cells are wired with the back contact of one cell connected to the back contact of the next. The total current is the sum of the individual current outputs of the cells, but the total voltage is the same as the voltage of a single cell. Cells are usually wired in series to form an array and arrays are wired in parallel to obtain desired current.

rectifier—A device that passes current in one direction only.

regulator—A device that prevents overcharging of batteries.

self-discharge rate—The rate at which a battery will discharge on standing; affected by temperature and battery design.

series—In array construction, connecting cells by joining the back contact of one cell to the front contact of the next cell to obtain a higher voltage.

shingling—In array construction, connecting cells by overlapping the front edge of one cell with the back edge of the next, similar to roof shingles, and soldering the edges together.

silicon—The second most abundant element in the earth's crust; intermediate-grade silicon (less costly than electronic-grade) is used in the manufacture of silicon solar cells.

solar cell—The photovoltaic device that converts sunlight directly into electricity.

state of charge—A battery's available capacity, stated as a percentage of rated capacity.

sulfation—A condition which afflicts unused and discharged batteries; large crystals of lead sulfate grow on the plate, instead of the usual tiny crystals, making the battery extremely difficult to recharge.

tracking system, 2-axis—A mount capable of pivoting both daily and seasonally to follow the sun.

tracking system, 1-axis—A mount pointing in one axis only, reoriented seasonally by hand and used with linear concentrators or flat plates.

V_{oc}—*see* **open-circuit voltage**

volt—Unit of electrical potential difference across which a current flows.

watt—Unit of power, power being the rate at which energy is used to do work.

INDEX

A practical application . . . .

[Photo: Photowatt International]